T0325560

Class A
ERP
Implementation

Integrating Lean and Six Sigma

BY
DONALD H. SHELDON
CFPIM, CIRM

APICS.
THE EDUCATIONAL SOCIETY
FOR RESOURCE MANAGEMENT

Copyright ©2005 by J. Ross Publishing, Inc.

ISBN 1-932159-34-7

Printed and bound in the U.S.A. Printed on acid-free paper
10 9 8 7 6 5 4 3 2 1

Library of Congress Cataloging-in-Publication Data

Sheldon, Donald H.
   Class A ERP implementation : integrating Lean and six sigma / by Donald H
Sheldon.
     p. cm.
   Includes bibliographical references and index.
   ISBN 1-932159-34-7 (hardback : alk. paper)
   1. Production planning. 2. Manufacturing resource planning. 3. Quality
control. 4. Six sigma (Quality control standard). 5. Organizational
effectiveness. I. Title: Class A enterprise resource planning
implementation. II. Title.
   TS155.S4445 2005
   658.5—dc22   2005001960

Direct all inquiries to J. Ross Publishing, Inc., 6501 Park of Commerce Blvd., Suite 200, Boca Raton, Florida 33487.

Phone: (561) 869-3900
Fax: (561) 892-0700
Web: www.jrosspub.com

DEDICATION

Special love and thanks to
Anita, Erica, and Geoff
for supporting me and my passion.
Special thanks to God for my many blessings.
"There but for the grace of God go I."

TABLE OF CONTENTS

PREFACE

This has been a fun book for me to write. The topic of Class A ERP is my passion, and those who know me know that I live and breathe this stuff. My master's thesis was written on Class A ERP, and my first book (with Michael Tincher) was also on the topic of Class A. That mid-1990s book was titled *The Road to Class A Manufacturing Resource Planning (MRP II)* (Buker, 1995), but a lot has changed since 1995. I have wanted to write the sequel for several years, but just had not had the opportunity or time. When J. Ross Publishing asked me to start it in the summer of 2004, it was a thrill. You may find it interesting that I used Class A practices throughout the writing of this book. I executed performance metrics and measured process to plan using measurements that included productivity, chapter length plan accuracy, and overall plan accuracy. I only hope that I have achieved Class A performance in the final product.

The thoughts in this book are ideas that have been proven by many companies and the people who run them. Some special experiences have ingrained process in my mind, and I have shared as many of these gems as possible. I am still learning and acknowledge that in a few years this book too will become obsolete, but for now I have faith that the focus areas, data accuracy, schedule accuracy, customer service, and cost avoidance will not go out of style soon. Technology plays an ever-increasing role in management systems and the pursuit of manufacturing excellence. I hope and trust that your journey will be enhanced by some idea in this book and wish you Godspeed!

ABOUT THE AUTHOR

Donald H. Sheldon is President of the DHSheldon & Associates consulting firm in New York. He started his career at The Raymond Corporation, a world-class manufacturer of material handling equipment. He held the position of Director and General Manager of Raymond's Worldwide Aftermarket Services Division when he left to accept the position of Vice President for Buker, Inc., of Chicago, a globally recognized leader in management education and consulting. While at Buker, Mr. Sheldon helped clients on every continent to achieve business excellence in numerous areas including inventory accuracy. After several years of traveling with Buker, Mr. Sheldon joined NCR Corporation, a client company, to work full time with its manufacturing facilities throughout Asia, northern Africa, Europe, and the Americas. As Vice President of Global Quality and Six Sigma Services, Mr. Sheldon was directly involved in the process improvement health worldwide at NCR. In 2003, Mr. Sheldon, to continue to support his passion for coaching excellence, launched DHSheldon & Associates. He and his network of consultants continue to work with companies in North America to improve competitive advantage.

Mr. Sheldon has published numerous articles in journals and is co-author (with Michael Tincher) of the book *The Road to Class A Manufacturing Resource Planning (MPR II),* published in 1995, and author of *Achieving Inventory Accuracy* (J. Ross Publishing, 2004). Both are available at www.amazon.com. He has been a frequent speaker at colleges, international conventions, and seminars including the American Production and Inventory Control Society (APICS). He holds a master of arts degree in business and government policies studies and an undergraduate degree in business and economics, both from the State University of New York, Empire State College. He is certified by APICS as CFPIM (Certified Fellow in Production and Inventory Management) and as CIRM (Certified in Resource Management).

Mr. Sheldon can be contacted at www.sheldoninc.com.

ACKNOWLEDGMENTS

The topic of Class A ERP is often misunderstood in the manufacturing environment. The chance to give the Class A ERP process its time in the spotlight was a great opportunity, and I thank J. Ross Publishing for it. Drew Gierman and the rest of the J. Ross staff have been great to work with, professional in every way, and extremely helpful on many fronts.

APICS has also been a great partner for years. I would like to thank the organization for many pleasant exposures. In my life, I have had the experience of serving on the local chapter board including as president, have been a member of the Southern Tier Chapter in Binghamton, New York for more than twenty-five years, have taught certification courses, and have spoken at dozens of meetings and seminars. In return, this organization has given me many opportunities.

Additionally, I need to thank Westy Bowen, master black belt at Textron and good friend, for being a Six Sigma resource while I was writing this book. He is truly a Six Sigma expert.

Lastly, I would also like to acknowledge a special friend with whom I traveled and team-taught many courses in my earlier days of consulting. Ed Turcotte, former friend and partner, was very helpful and understanding as I learned to spread my wings in this field years ago. I am only sad he had to leave to be with the Lord so early in his life. We had much more work to do together. Class A made him smile as well. I hope he is still smiling as he looks down.

ABOUT APICS

APICS — The Educational Society for Resource Management is a not-for-profit international educational organization recognized as the global leader and premier provider of resource management education and information. APICS is respected throughout the world for its education and professional certification programs. With more than 60,000 individual and corporate members in 20,000 companies worldwide, APICS is dedicated to providing education to improve an organization's bottom line. No matter what your title or need, by tapping into the APICS community you will find the education necessary for success.

APICS is recognized globally as:

- The source of knowledge and expertise for manufacturing and service industries across the entire supply chain
- The leading provider of high-quality, cutting-edge educational programs that advance organizational success in a changing, competitive marketplace
- A successful developer of two internationally recognized certification programs, Certified in Production and Inventory Management (CPIM) and Certified in Integrated Resource Management (CIRM)
- A source of solutions, support, and networking for manufacturing and service professionals

For more information about APICS programs, services, or membership, visit www.apics.org or contact APICS Customer Support at (800) 444-2742 or (703) 354-8851.

*Free value-added materials available from
the Download Resource Center at www.jrosspub.com*

At J. Ross Publishing we are committed to providing today's professional with practical, hands-on tools that enhance the learning experience and give readers an opportunity to apply what they have learned. That is why we offer free ancillary materials available for download on this book and all participating Web Added Value™ publications. These online resources may include interactive versions of material that appears in the book or supplemental templates, worksheets, models, plans, case studies, proposals, spreadsheets and assessment tools, among other things. Whenever you see the WAV™ symbol in any of our publications, it means bonus materials accompany the book and are available from the Web Added Value Download Resource Center at www.jrosspub.com.

Downloads available for *Class A ERP Implementation: Integrating Lean and Six Sigma* consist of more than one hundred PowerPoint slides to help support a successful Class A ERP implementation.

WHAT IS ERP?

Process improvement and the need for it move fast in today's competitive markets. To stay on top of the latest implementation successes and techniques, it would seem that one must read continuously. New "buzzwords" are invented every week by some organization trying to sell its wares. Some new acronyms stick and some do not. I am in the business of sorting this stuff out and have come to believe that *implementation and success* is the final test for validity. I have been in the implementation business since the mid-1980s, and in that time I have determined that some of these process methodologies are actually legitimate. Three specifically are (1) Class A ERP, (2) lean, and (3) Six Sigma.

I have also observed through working at manufacturing and distribution facilities all over the world that there are some specific high-value factors embedded in each of these approaches. After several years of involvement, observation, and even, admittedly, some trial and error, I have come to the conclusion that there is definitely a sequence of events and an approach that work best for successfully harvesting the goodness with the least amount of disruption and best competitive advantage. That is what this book will focus on — how the improvement processes integrate and the best approaches to implement them. In reality, the method is not the goal; the goal is implementing improvement and sustainable good habits.

> In reality, the method is not the goal; the goal is implementing improvement and sustainable good habits.

Some readers of this book will have a burning desire to get right to the integration discussion — how to best integrate enterprise resource planning (ERP) with lean and Six Sigma. Be assured we will do that with plenty of detail and real-life examples as this book progresses, but right now the foundation we must start with is a common understanding of ERP: what it is and why organizations would want to do it.

There is a lot of confusion about the topic of ERP. This confusion is often credited to the aggressiveness of business system software peddlers. Not that ERP tools do not offer a lot of value-add, because they clearly do. Tiger Woods would have had a difficult time maintaining a top ranking in the world of golf if he did not have quality, straight, and true golf clubs and perfectly balanced golf balls to limit the process variation. The rest of the story is meaningful as well, however. Without his skillful and disciplined execution, the tools alone would not get the job done to a high-performance standard.

Like a professional-level game of golf, high-performance manufacturing requires discipline and execution with the highest process repeatability and predictability. ERP is a business model that involves all levels of the organization — hence the word "enterprise." ERP process disciplines allow organizations to link customers and top-management decisions all the way through to execution in the supply chain and the factory floor. Well-executed ERP not only starts with top management, it is totally *dependent* on top management.

> Well-executed ERP not only starts with top management, it is totally *dependent* on top management.

At this point, I should admit to you that I have been involved with a few Class A ERP implementation *failures* as well as successes. It needs to be said that the common denominator in the failure of any of these improvement process approaches is always the same thing — lack of top-management involvement and prioritization of activity. That is a very well-known fact. Let us look at how high-performance organizations handle top-management planning.

TOP-MANAGEMENT PLANNING

Top-management planning is clearly the most important process within ERP. As illustrated in Figure 1.1, it is clear that many decisions must be made to guide the organization's vision correctly. While the "strategic" element of top-management planning may be the single most effective component of ERP, the rules and *spirit* of ERP do not dictate strategic policy, but instead insist on strategic linkage to the rest of the organizational practices and execution, especially the demand-side and supply-side rules of engagement. More simply stated, the top-management ideas must actually drive activity in the organization. There needs to be a direct connection.

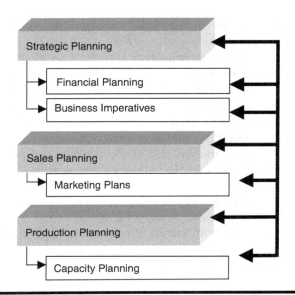

Figure 1.1.   Top-Management Planning.

ERP is the spinal cord and information flow that link top-management thinking and planning with marketing, sales, capacity planning, procurement, manufacturing, and customer service. The top-management planning engine is worthless if there is no transmission to link top-management thinking directly with the execution of those plans.

The top-management planning engine is worthless if there is no transmission to link top-management thinking directly with the execution of those plans.

I remember a lesson from a management book I read many years ago. I found the lesson amusing and it has stuck with me all these years. The saying went something to the effect of "Business planning in many companies is too often like a rain dance. Managers get together and ceremoniously dance and make a lot of noise, but at the end of the day, it doesn't have much to do with the weather." I can relate to this humorous story directly from my observations and travels to many companies. I go into a lot of businesses where top-management teams do exactly that — have a lot of fun and make a lot of noise about strategic plans, but at the end of the day, this activity does not have much to do with the execution of plans. When executed in its proper form, ERP connects these processes. It creates the linkage and management systems to ensure proper execution of marketing plans, sales policy, supply chain partnerships, and capital spending. When executed properly, ERP does this through a formal management system and, to some degree, supporting tools.

In a manufacturing business, management systems are planned and executed accountability infrastructures that create predictable opportunities for follow-up on decisions and goals. Examples would include infrastructure events such as daily performance reviews, weekly project management reviews, top-management monthly planning performance reviews (often referred to as sales and operations planning [S&OP]), and events directly related to master scheduling, such as a "clear-to-build" process. All of these events and processes such as master scheduling and the clear-to-build infrastructure will be described in detail in the coming pages.

Class A ERP, which we will talk about in the next chapter, *dictates* many elements of performance review. Along with agreed-to and documented "rules of engagement," these result in increased predictable performance. In Chapter 6, the details of how this top-management planning process works in high-performance businesses will be described.

MASTER PRODUCTION SCHEDULING AND MATERIALS MANAGEMENT

The outputs from top-management planning feed directly into the master production scheduling (MPS) and materials management when executing correctly. Figure 1.2 illustrates the subsequent activities.

While master scheduling in many businesses is a clerical process of documenting the promises made by order management, when done properly it is much more. Master scheduling is anything but a simple process. Master scheduling done correctly is a science *and* an art.

> Master scheduling done correctly is a science *and* an art.

The truth is, if you want to be influential in any manufacturing facility, the position you want is either plant manager or master scheduler. In Chapter 7, the rules of master scheduling and the roles of the master scheduler will be reviewed in detail. To spike your interest, here are some introductory thoughts on the function and responsibilities of good master scheduling within ERP.

The short story on a high-performance MPS process starts with the idea that top-management planning (known as S&OP) drives product family demand and operations expectations in both supply-side and demand-side requirements. S&OP has been growing in popularity for the last several years and has become a key and critical component of both ERP and good management process in general.

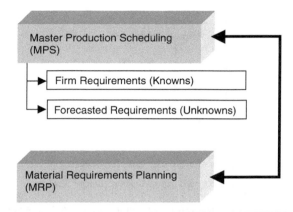

Figure 1.2. Master Scheduling.

The S&OP process outputs are made up of "product family" requirements. It is at this level of detail where the MPS starts to add some real value. In most environments, "product families" without any more detail than that cannot be manufactured without more specific information. These product family signals are translated directly into detailed, firm, and forecasted signals by the MPS through planning techniques. When the ERP process works efficiently and effectively, scheduling becomes the linkage rod or transmission connecting the top-management planning engine with the execution wing of the business, normally operations.

> When the ERP process works efficiently and effectively, scheduling becomes the linkage rod or transmission connecting the top-management planning engine with the execution wing of the business, normally operations.

When the ERP process is *not* working properly, the linkage does not exist. The scheduling process becomes a stand-alone process reacting to daily demand with short planning horizons. The results can range from ineffectiveness to devastation, and this rarely results in long-term success. In today's best-managed businesses, few choose not to hold the S&OP process as a valuable component of their top-level management system. As Tim Frank, CEO of Grafco PET Packaging Corporation, said to me, "Why wouldn't any top manager want to do this?"

Master scheduling takes the signals from top management and translates them into usable requirements for production planning. It is the master scheduler's job to determine the mix within product families and other unknowns such as lot size and priority of order requirements. Without these important communi-

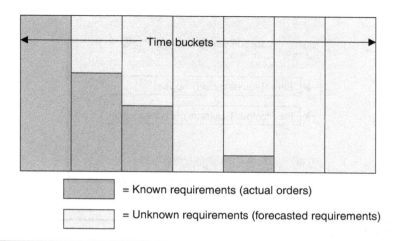

= Known requirements (actual orders)

= Unknown requirements (forecasted requirements)

Figure 1.3.   Scheduling "Knowns" and "Unknowns."

cations, the planning engine, material requirements planning (MRP), would not have anything to plan, or if it did, there would probably not be many accurate requirements posted other than actual orders. That is due to the fact that much of what the master schedule sequences in the schedule are "unknowns" (see Figure 1.3).

Unknowns are often the drivers of unnecessary inventory buffer, so it is critically important to take this scheduling process seriously. The best master schedulers work with their organizations' supply chain managers to minimize lead time and increase flexibility within the supply base and, at the same time, work with the order management team to establish responsive rules of engagement that put everybody on common ground in terms of both goals and expectations. The master scheduler's job is much like the conductor of an orchestra. The musical score is customer need, and the master scheduler interprets customer need and translates it into the schedule. There are several players in manufacturing and supply chain management, just as there are in an orchestra, and they all need to be coordinated. The baton's drumbeat is the master schedule itself broadcasted to the team. When everybody is exactly in sync at the end of the song, there is no extra unplanned inventory buffer and the music was done exactly to the composer's liking.

There are several inputs to the MPS, and the more accurate these data are, the more likely the process will be successful and value-add. For this reason, Class A ERP performance also includes data accuracy as a specific area of focus.

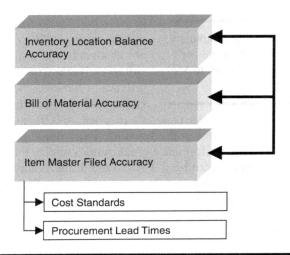

Figure 1.4.   Database Accuracy.

DATA ACCURACY

ERP inputs play an important role in the effectiveness of the overall ERP process. Data accuracy of these inputs obviously plays a critical role, and data integrity is a historically proven prerequisite to high performance. Accuracy of data is an asset for process improvement as well as process predictability in all high-performance organizations. Data accuracy elements in an ERP focus always include inventory balance accuracy, bill of material (BOM) accuracy, and other item master file field data accuracy elements, such as lead time and cost standards (Figure 1.4). Since the ERP process focuses on data and process linkage, it becomes extremely important for the data elements to have high integrity. When inputs to the ERP process such as inventory balances (which directly affect MRP) are accurate, the user has a significantly improved chance of producing valuable and accurate outputs such as supply chain signals and shop floor orders. Examples of these accuracy-dependent processes include not only requirements planning but also cost planning and inventory management.

BILL OF MATERIAL ACCURACY

The discussion on BOM accuracy is quite possibly of even more importance than the inventory balance integrity. BOMs are the recipe that is used by many organizational functions. This would include kits in distribution and assemblies,

If BOM information is not absolutely right, it becomes difficult to deliver a cost-effective, quality product to the customer.

batch mixes, weldments, and kits in manufacturing. If BOM information is not absolutely right, it becomes difficult to deliver a cost-effective, quality product to the customer. The product either has wrong components and informal process to get the correct materials in place, specifications are wrong, or it is reworked before it leaves the building. Regardless, it is unnecessary process variation. All high-performance organizations realize the importance of these data and treat them accordingly. Few would argue against the value-add characteristic of accuracy in the building of requirements for engineering design, material requirements, quality standards, and cost planning. It is important not only that the BOM record within the ERP business system meet the engineering specification, but also that these records match what the factory floor actually does!

INVENTORY BALANCE ACCURACY

Inventory accuracy is another key element of ERP data management. Inventory records are the jewels of the planning process and affect both requirements planning and costs. In the planning process, there are two types of inventory records: on hand and on order. Both need to be accurate for the planning process to work properly. Like all topics covered in this first overview chapter, this too will be given ample attention later in the book as each element of ERP is detailed. Figure 1.5 illustrates the interaction of data records with the planning engine of ERP, master scheduling, and MRP.

The execution processes of ERP are also outputs of the planning process and are literally the end of the trail where ERP outputs from plans at all levels are executed (see Figure 1.6). These include procurement, shop floor controls, product delivery, and service execution. Supply chain management is controlled from this vantage point. Without these signals, the supply chain cannot be linked solidly to the manufacturer's vision and drumbeat. In many manufacturing organizations, purchasing functions under confident but less informed management act quite independently and do not have tight linkage to the MPS process. In high-performance ERP organizations, procurement processes are tied directly to the ERP planning processes just as tightly as other ERP outputs. The benefits of this linkage discipline become obvious.

Inventory is a direct result of purchases. Buying the wrong inventory means wasted energy and money on several levels. Buying too much inventory is

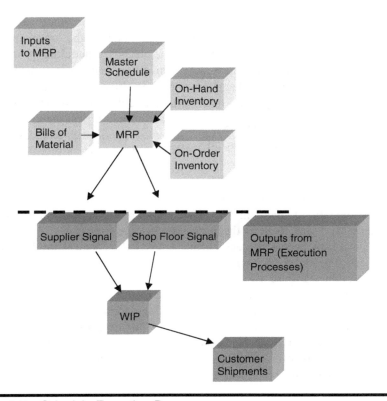

Figure 1.5.   Schedule Execution Processes.

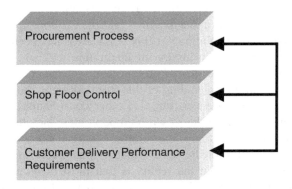

Figure 1.6.   ERP Execution Processes.

especially wasteful in a cost-conscious competitive world. Only disciplined process linkages within the information flow initiated by the top-management planning process and scheduled through the master scheduling process can allow maximum efficiency and accuracy of inventory-buy commitments. Without the direct information linkage, inventory is driven from best pricing or truckload quantities only and not necessarily strategic amounts. This *can* result in additional obsolescence, stranded inventory, and other unnecessary cost drivers. Inventory is often referred to as an evil, but it is quite the contrary.

> Inventory is often referred to as an evil, but it is quite the contrary.

Inventory is the best asset you can have at the time the customer comes to buy it! ERP allows the opportunity for this synchronization to happen. As we will find throughout this book, discipline is the necessary glue to keep the processes linkage intact.

You have now been introduced to the entire ERP business model, but only in its pieces. The entire ERP business model in high-performance organizations usually resembles Figure 1.7.

In the model, one can see the importance of information flow and process linkage from top-management planning through to plan execution and, ultimately, customer service and satisfaction. In Chapter 2, the discussion around specific performance requirements prescribed in Class A ERP certification will begin. Certification offers some value when shifting culture within the entire organization as it creates a specific performance target and celebration point.

INTEGRATION WITH LEAN AND SIX SIGMA

In this book, we will first dissect the ERP business model, convert the process elements into Class A ERP understanding, and then follow up by showing the integration to lean and Six Sigma. At this point, ERP has been described with enough detail to begin the discussion of integration.

Many organizations have *started* on journeys to excellence, yet a high percentage fail. I have an understanding of this from not only clients, practitioners I meet, and personal experiences, but also other consultants I meet. Although these consultants often have their own methodologies, inevitably we find that we are all doing the same thing, but simply calling it by a different name.

After all, we are all after the same result — manufacturing excellence and competitive advantage — but the reality is that most of us are simply trying to avoid the wrong way. It is the big mistakes that add unnecessary time and cost. From my experience, there *is* a way to implement a high-performance or world-class process that works effectively and efficiently every time. To do so,

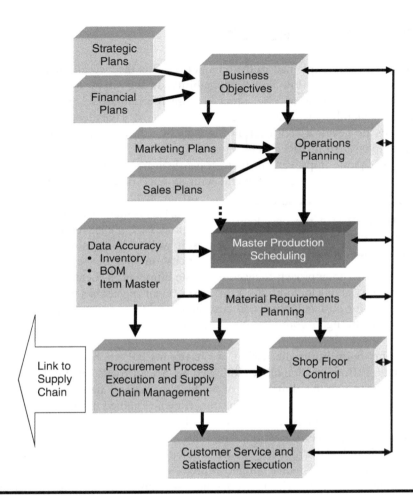

Figure 1.7. ERP Business System Model.

it is best to understand this journey for its pieces. This book's goal is to define and separate the components of the journey most referred to as ERP, lean, and Six Sigma. No other initiatives (for the foreseeable future) are necessary, and it will be more understandable by the time you get to the last chapter of this book. It all comes back to the discussion of improvement implementation methodology and what we choose to call it. ERP, lean, and Six Sigma are the names we will call the segments of our journey to world-class performance. You can probably think of other names that some of the elements we are going to discuss have been called.

ERP AS THE STARTING POINT

As I have briefly explained in this chapter, ERP is the overall business model defining information flow and accountability. In the simplest form, high-performance applications of ERP processes are repeatable and predictable. The focus is generally on data and schedule accuracy. That does not mean that processes are necessarily wasteful or low quality; they just are not normally the main emphasis from ERP itself.

As a separate improvement focus topic, lean normally would not reference a business model such as we would with an ERP discussion. Lean is from a different space. Whereas ERP is a process, lean is more of an approach within a process. For that reason, it lends itself well to be a completely separate focus on the journey to excellence.

> Whereas ERP is a process, lean is more of an approach within a process.

Lean refers to an approach focused on waste elimination. Lean thinking within a business is about looking at all processes, even repeatable processes, as opportunities for cost reduction and customer service improvement. This is normally done through rigorous process evaluations using mapping and other problem-solving tools.

At this point in the discussion, we have an ERP emphasis on process and information flow predictability, followed by a lean focus on improving these processes through the elimination of waste. Nothing new here for most readers, but you may begin to get a sense for the efficiency gained through the sequence of focus. That is exactly the same sense I have developed by helping hundreds of businesses all over the world. The issues are all the same in the businesses I have visited. Get the processes to work quickly and then come back to improve them through a new focus. Shigeo Shingo, the famous Toyota improvement wizard, used this approach to improvement — plan levels of improvement from the beginning, improve by 50 percent, and then come back to improve by 50 percent again!

At what point does Six Sigma come into this discussion? Many managers have had these thoughts when reading about Six Sigma and relating back to their statistical process control experiences of the 1980s. "How the heck are complex statistics with a goal of zero defects going to help real-world problems on the shop floor? Those Six Sigma guys live in a dream world." The sad truth is that many organizations start their serious improvement effort with Six Sigma after the CEO reads an article on an airplane. His or her epiphany is then followed by education and six to twelve months of floundering and making projects (that were already in process pre-Six Sigma) fit the new Six Sigma requirements. Six Sigma as a quality goal can be reached through tinkering as well, if that label alone is the goal. After all, defects *are* defined by where the lines are drawn around customer requirements. Anyone who concentrates on describing the requirements to your capabilities can achieve Six Sigma — well, sort of — in name only, certainly not the spirit of Six Sigma. Many of these companies end up seeking additional help. Having *been that help* in some of these organizations, I have found that the best approach starts with understanding the basics. Call them what you want, but these basics are back to the ERP discussion. Six Sigma is a sophisticated vision and organization of problem-solving and process-improving teams and tools. Once an organization is ready for this level of cultural focus for tenacious improvement, Six Sigma is the right methodology. The reality is that it takes organizational maturity for this to work. Many *successful* organizations have taken the improvement process in progressive steps.

> Once an organization is ready for this level of cultural focus for tenacious improvement, Six Sigma is the right methodology.

Six Sigma is a powerful set of tools and a methodology that can empower a disciplined and trained organization to go from good performance to great. Six Sigma is literally a designation of quality as defined by standard deviations. That is as far as we are going in this chapter. There will be much more to come in following chapters.

As illustrated in Figure 1.8, when you look at each of these improvement methods for its specialty and strength, you find that they complement each other quite well. In every company I have been involved with, I have found this to be applicable and a valuable understanding to start with. Having a good set of ERP standard practices that predictably result in promises met is a great foundation on which to build a culture of continuous improvement. The logical next

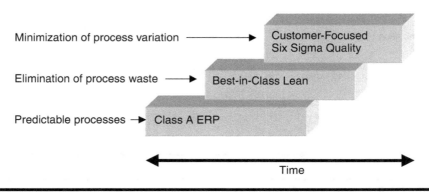

Figure 1.8. Sequence of Focus.

step in many organizations is not to jump into complex statistical tools and deal with all of the complexities of hours and hours of in-depth training required to establish Six Sigma green belts and black belts. While there is a time and place, I have found it to be more efficient to keep the process focused on significant yet incremental improvements with a clear vision of where you are headed, including the expectation of Six Sigma levels of quality and process reliability. Be acutely aware that I am not downplaying the importance of Six Sigma process methodology. Quite the contrary; I found it to be a very powerful tool, and when organizations, including management, are ready for this level of focus, it is the right step to take.

The final step in this vision is world-class performance (Figure 1.9). I am sure there are more steps to come and I hope I live long enough to see them, but it is difficult to see beyond world-class performance. Most would agree that it is a little foggy on that level. World-class performance is not yet clearly defined. This performance level also changes every day, and to make matters even more confusing, it is not defined by the same market week to week or process to process; new product introduction may be defined by the electronics consumer goods market, lean flow may be defined by the automobile industry, quality may be defined by the aerospace market. This last step is not defined in this book because it just is not thought through enough in the marketplace and there are too few organizations that have reached that level to prove the theories.

I can only imagine that technology will continue to play an expanding role in business. Who knows? RFID (radio frequency identification device) is al-

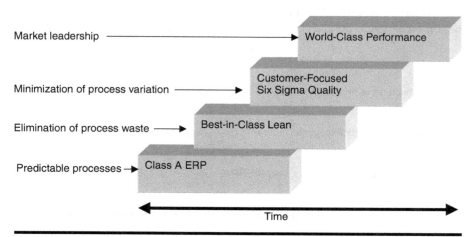

Figure 1.9. Journey to World-Class Performance.

ready starting to change the manufacturing and distribution landscape, and I can only imagine the world in twenty years. World-class manufacturing and a *real* definition for it — that is a good topic for a future book!

At this point, I hope that you are starting to understand the interdependencies in the prerequisite steps to world-class performance. Class A ERP performance is a major step in the right direction in every business that focuses on it. The management system requirements from Class A ERP greatly help the organization mature in terms of both accountability and discipline. It is time to take a more in-depth look at Class A ERP.

planning process and a major component of the Class A MRP II and ERP requirements. Top decision makers in the business could now influence activity more directly at the plant level, including inventory levels and general risk management involving capacity and anticipated demand.

At this point, some readers may be hungry for more Class A MRP II criteria detail. Your hunger would not have been satisfied in this era, however. The measurements were fairly loosely defined and could be widely different from business to business and one consultant to another. Nonetheless, Class A MRP II was quite popular as word got around about the benefits companies were seeing as a result. The early to mid-1980s period was the heyday for MRP II consultants. David W. Buker, Inc., of Antioch, Illinois, was named during this period as one of the fastest-growing companies in America by a leading business magazine. Oli Wight first and, later, David Buker developed popular video education in this space that helped deal with the growing need for education and training of the masses within the manufacturing world. These videos sold for thousands of dollars based on the return that many businesses enjoyed. Additionally, public classes on the topic delivered by the masters were filled regularly, and businesses in need of a competitive advantage set out to implement this new formula for success. Needless to say, some companies were much more successful than others. It depended on the approach and discipline. In most cases, there were still gains. Even with this rudimentary approach, because the general population of manufacturing businesses at the time had extremely poor discipline in scheduling and data accuracy, even some half-hearted attempts at Class A MRP II implementation yielded benefits. The benefits commonly experienced included reduced inventory, increased customer service, higher productivity, and lower costs.

At the time, implementation lead times (time from initial Class A MRP II education to certification) often exceeded eighteen to twenty-four months. In the 1980s, data were not as readily available as they are today. That lack of underlying data and facts, as well as the associated root causes, certainly left some opportunities on the table. The first implementation I was involved in went quite well according to the standards of that time, but I am embarrassed to tell you that it took twenty-six months to complete certification. We hit Class A standards, but were not able to show consistent ability to sustain the performance until the second year. Today, a twenty-six-month implementation would be considered a disaster!

The business environment continued to change and the competitive bar was raised in most markets. MRP II expanded and grew as new process successes were tried. This evolution came from people exchanging ideas. People in this field are in the business of idea agriculture. As the late Ed Turcotte (a former consultant partner and friend of mine) used to say, "If we see great ideas, we harvest them and then plant seeds in other factories." This is the business of consulting. Take ideas that work and spread them. That is exactly how Class A evolved from the MRP skeleton. It is also why ERP will continue to evolve in the future.

The 1990s

In the 1990s, especially the mid-1990s, the Class A model evolved from a set of metrics to a process with several required components. Influences from various successful implementations as well as from other evolving methodologies, including ISO (International Standards Organization), total quality, and even to some degree just in time, all influenced the process.

ISO, which in the early days had no more value than to fill a political gap in trade agreements between the newly formed European Union countries, even contributed in some small way as documentation of process gained more respect in securing repeatability through process control. During this time, thought leadership within Class A teaching developed both a management infrastructure for accountability and an additional business model structure. The first thoughts that were centered around a progression of objectives within three areas of focus — process design, management system, and results — started to affect behaviors and aid in process control.

During the early to mid-1990s, the implementation times of Class A performance and structure started decreasing even though the certification requirements were raising the bar on performance. Most organizations that were serious about their implementations were able to reach Class A performance levels successfully within twelve to eighteen months. The efficiencies came from top management being more involved and follow-up being more predictable and planned. When process owners take the time to do root cause analysis and drive actions, this kind of improvement can happen fast! Most organizations were also finding that education, the investment in human capital, was often worth the monetary capital when topics were carefully chosen. This was especially relevant in organizations that were upgrading their software tools.

Software companies had a growth period in the 1990s as computer technology and process evolved quickly. It was during this period that the acronym and term "ERP/enterprise resource planning" was frequently marketed to differentiate one software package from another. ERP became popular, and soon the term MRP II became "yesterday's newspaper." The process requirements for Class A performance continued to evolve, and the new name for it was ERP. It had a more modern ring to it.

Because the scope of Class A performance was growing with the competitive pressure of most manufacturing markets and admittedly because the market was asking for ERP rather than MRP II, Class A upgraded its cloak to Class A ERP as well. As the competitive pressures increased from global markets, so did the need for process improvements in manufacturing. In the late 1990s, Class A ERP became quite refined, especially as compared to the early days of MRP. By then, the Class A process had elements of expectation not only in scheduling and planning but also in inventory strategy, quality, demand forecasting, supply chain management, project management, and plan execution. The Class A ERP management system had grown into monthly, weekly, and daily elements and expectations. Not only were monthly top-management planning events scheduled well in advance, but there were weekly and daily regimens required by high-performance organizations. A predictable accountability infrastructure was scheduled in six-month intervals as well as twelve-month schedules, all dictated by the evolving Class A ERP process criteria. Some of the management system events are listed in Table 2.2.

In each case, these management system events helped to control the speed and sustainability of conversion to Class A ERP performance. Class A ERP was becoming much more predictable, company to company and consultant to con-

Table 2.2.   Management System Elements

Time Frame	Process Event	Process Expectations
Yearly	Strategic review	Updates to the strategic plans
	Business imperatives	Prioritizing the short list of "must do" objectives
	Talent review	Management assessment of key employee skills
	Succession planning	Key position and skills analysis of bench depth
Monthly	S&OP	Risk management of capacity, inventory, and customer service decisions
	Project review	Review progress on business imperatives
Weekly	Performance review	Process owner review of progress
	Project review	Detail review of projects by process owners
Daily	Schedule review	Detail review of yesterday and today's requirements
	Daily walk-through	Management-by-observation tour of the factory

sultant, even though there was no world standard such as for ISO or Six Sigma. Implementations in some instances were now, for the first time, taking no more than six months from initial Class A ERP education to certification.

In the mid-1990s, I was working with AlliedSignal (now Honeywell). The company was in the process of converting to SAP business management software. It was a vision of the management at AlliedSignal to have the disciplines in place prior to the implementation of new software. They had the insight to use Class A as the standard. At the time, there was enough standardization accepted within Class A that three separate consulting companies were approved as sources for this training and certification. Buker, Inc. (where I was a partner) was one of them. The management vision was communicated to the plants at the time; it stated simply that the new software would not be available except at facilities that had met Class A criteria to a certain level. This proved to be a good prerequisite for software performance. This effort resulted in significant improvements in inventory control and accuracy, schedule disciplines and execution, and ultimately customer service.

Topics like transaction design were addressed and any process bugs worked out prior to turning on the new software. Data were measured, scrubbed, and maintained through Class A management systems. It took those process variation areas that are common in many organizations off the list of software implementation issues.

The new ERP acronym was emerging as the replacement for MRP II, and it made a lot of sense. The message was that supply chain management was much bigger than the MRP process. The use of the word "enterprise" suggested a much bigger planning scope. It was time for a change. ERP had the right mix of new ideas and proven processes to be the successor to MRP II.

> I, for one, never liked the fact that MRP II had an acronym similar to the simple process calculation of MRP.

I, for one, never liked the fact that MRP II had an acronym similar to the simple process calculation of MRP. It has always caused confusion. As MRP II changed to ERP, I said "good riddance!"

THE CLASS A ERP STANDARD OF TODAY

Because this book is about the Class A ERP standard and because it will be detailed in coming pages, it is probably best simply to describe the main principles at this point. This will give the reader the place from which I start the discussion of Class A ERP. Boiled down, Class A ERP is about proficiency in all of the areas of focus, as shown in Table 2.3.

Table 2.3.   Class A Focus Areas

Prioritization and Management of Business Objectives

1. Project management
   a. Project funnel
   b. Prioritization of projects
   c. Resources and skills required
   d. Review process
2. Human capital management and investment
   a. Professional society affiliations
   b. In-house education
   c. Tuition aid programs and guidelines
   d. Training
      i. New employees training
      ii. Existing workforce training
      iii. Skills assessment
3. Business imperatives
   a. Hoshin planning
   b. Review process and documentation
4. New product introduction
5. Accountability infrastructure
   a. Metrics
   b. Management systems
      i. Daily
      ii. Weekly
      iii. Monthly

S&OP Processes

1. Strategic planning
   a. Markets
   b. Core competence
2. Demand planning
   a. Mix
   b. Volume
3. Operations planning
   a. Supply chain partnerships
   b. Capacity planning
      i. Internal capacity
      ii. External capacity
4. Financial planning
   a. Profit
   b. Capital spending
   c. Revenue

Scheduling Disciplines and Production Planning

1. Master scheduling
2. Rules of engagement

Data Integrity

1. Inventory location balance accuracy
   a. Warehouse design
   b. Transaction design
   c. Point-of-use storage
   d. Location design
      i. Raw
      ii. Components
      iii. Work in process
      iv. Finished goods
   e. Cycle count process
      i. ABC stratification
      ii. Tolerances allowed
2. BOMs or bills of resource accuracy
   a. Engineering change
   b. Process to repair BOMs
   c. Audit process
   d. Routing linkage to BOM
3. Item master accuracy
   a. Lead times
   b. Cost standards
4. System security
5. Part number design

Execution of Schedules and Plans

1. Procurement process
   a. Linkage to MPS
   b. Supply chain communications process
   c. Management systems
2. Shop floor control
   a. Linkage to MPS
   b. Communications process
   c. Management systems

The measurements listed previously got a lift to new standards, and the metrics in the 2000s appear in Table 2.4.

Not only has the number of metrics increased, but so has the threshold of acceptability. These metrics are built from an ERP business model that shows

Table 2.4.   2000s Class A Metrics

Measurement	Minimum Performance Requirement
Profit and/or budget accuracy	95 percent
Sales forecast accuracy by product family	90 percent
Production plan accuracy by product family	95 percent
MPS accuracy	95 percent
Safety	0 recordable accidents
Schedule stability	95 percent
First-time quality	97 percent
Inventory record accuracy	98 percent
BOM accuracy	99 percent
Item master accuracy	95 percent
Daily schedule adherence	95 percent
Procurement process accuracy	95 percent
Customer promise accuracy	95 percent
Overall performance	95 percent*

* Sustainability normally is required and proven by showing 95 percent performance overall for at least three months.

succinct linkage between levels and activities, the essence of Class A ERP. Class A ERP is not about the metrics. It is much more than that. It is about the process design and intent, the management systems to ensure the sustainability and improvement of the process execution, and the results or performance. The results are evidenced by the metrics.

In Figure 2.1, you see the whole manufacturing organization, from top-management planning to shop floor execution. The large arrow pointing left from the procurement process indicates where the link would exist between this model and the identical ERP model in the supplier's business. Keep in mind that this model is applicable in any business, from process flow company to sheet metal shop. I have personally coached the implementation of this Class A ERP business model in business markets including plastics, appliances, auto parts, sporting goods, medical equipment, baking, construction equipment, capital goods, engineered specialties, paper goods, electronic consumer goods, harness manufacturing, electronic circuit boards, and many others. Class A ERP will work in not only your business, but also your supplier's.

Class A ERP will work in not only your business, but also your supplier's.

Having the ERP business models linked is the essence of supply chain management. Acknowledging the rules, roles, accountability, and information flow is a major element of Class A ERP performance. The gains come from disciplines of process as well as shared goals.

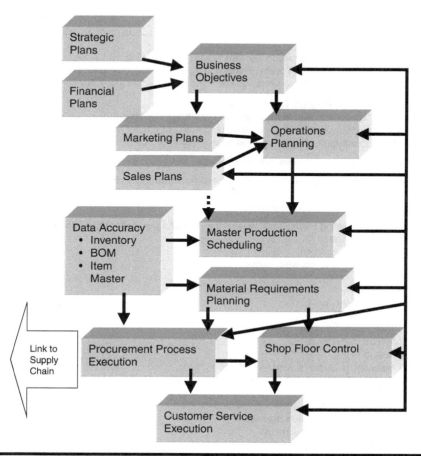

Figure 2.1. ERP Business System Model.

CLASS A ERP CERTIFICATION

Certification is a confirmation from an outside party that you have met the rigid criteria of Class A ERP performance. The criteria are areas that correspond with Table 2.3. A certification document that is several pages long is administered by an experienced auditor. Companies will often prefer to have an audit done a month or so prior to the final one to get a "gap analysis" punch list. Since Class A is so all-encompassing, this often makes a lot of sense. At DHSheldon & Associates, we subscribe to the audit agenda shown in Table 2.5.

The last items for discussion within certification are the focus areas around each topic. During the certification, the auditor is looking for three different

Table 2.5   Class A ERP Audit Agenda

Walk-Around
1. Review 5-S (Toyota Production System term) housekeeping and workplace organization
2. Observe "visible factory" (Toyota Production System term) communication boards
3. Discuss daily performance with a worker randomly selected from the floor

Review of Class A–Related Documentation
1. Check for key points important to the sustainability of Class A
2. Check for good process design, especially in areas of master scheduling, S&OP, inventory accuracy, management systems, and measurement processes

Observe the Top-Management S&OP Process

Discuss and Observe Evidence of the Strategic Planning Process
1. Review documents
2. Observe the business imperative list

Spend Time Reviewing Each Process with the Process Owner for the Following Processes
1. Master scheduling
   a. Rules of engagement
      i. Practices and communication
      ii. Time fences
   b. Disciplines
   c. Measurement audit
2. Materials planning
   a. Rules of engagement
   b. Measurement audit
3. Purchasing
   a. Check for past-due purchase orders
   b. Check for weekly maintenance of open purchase orders
   c. Review metric
   d. Review supply chain relationships and partnership agreements
   e. Review reverse auction process
4. Shop floor schedule attainment
   a. Review metric
   b. Review process for assigning daily schedule
   c. Check for linkage to the MPS
5. Education and training
   a. Review policy and practices
   b. Review documentation of employee education and training
   c. Review skills assessment process
   d. Discuss involvement with professional societies and review policy regarding same
   e. Review policies for tuition reimbursement and educational support for off-campus education
   f. Review new employee training policies and execution of same
6. Project management process
   a. Review project funnel process
   b. Review prioritization process
   c. Audit top manager (within the facility being audited) project review process
   d. Understand project linkage to business imperatives of the business
7. Quality
   a. Review first-time quality metric

Table 2.5   Class A ERP Audit Agenda (continued)

8. BOM accuracy
   a. Review metric
      i.   Specification
      ii. ERP business system record
      iii. What workers do on the factory floor
   b. Review engineering change practices
   c. Audit average cycle time for making changes to inaccurate BOMs discovered by the business

Customer Promise Accuracy
1. Review promise process
2. Review metric

Management System — Accountability Infrastructure
1. Review documentation
2. Observe weekly performance review
3. Observe daily walk-through or daily schedule adherence meeting
4. Observe weekly clear-to-build process
5. Observe project management review

New Product Introduction Process
1. Review process gates
2. See evidence of disciplines
3. Review past postlaunch audits done internally for past product introductions

Inventory Accuracy Review
1. Review transaction processes
2. Review cycle count process methodology
3. Audit raw and components storage accuracy
   a. Take sample counts and check accuracy to the system's perpetual record
   b. Review cycle count sheets from recent factory counts for process integrity
4. Audit finished goods accuracy
   a. Take sample counts and check accuracy to the system's perpetual record
   b. Review cycle count sheets from recent factory counts for process integrity
5. Review work-in-process accuracy management system
   a. Check definitions and applications of "controlled inventory areas"
   b. Check for the proper application of the "rule of twenty-four." (The "rule of twenty-four" has to do with the length of time in hours that a stockkeeping unit [SKU] is planned or by practice kept in one area at a time. If it is more than twenty-four hours, the perpetual record in the ERP system should be updated as a regular practice at the time of the physical inventory transfer.)

aspects of each topic: (1) robust process design, (2) management systems to make sure it is happening and will continue to happen, and (3) results that are synonymous with the metrics.

In addition to these three areas of focus, there is the concern of sustainability. Organizations must prove that they can sustain Class A ERP levels of performance. This is usually proven by three months of sustained results or clear trends showing consistent improvement in the results over time.

Process Design Review

In the process design review, the auditor is looking for proper and acceptable procedures evidenced by good documentation. An example of noncompliance to Class A ERP criteria would be if the organization did not have proper transaction documentation for inventory transactions. Another noncompliance example might be if the transaction design included too many transactions or was too complex for easy training and execution.

Management Systems Review

The second level of audit is the management systems review. The focus here is to ensure that the processes are set for sustainability. Normally, that means that there is some "fool proofing," "go, no-go" check, or accountability infrastructure. Continuing to use the inventory transaction example for discussion, a noncompliance example could consist of a well-designed transaction process for inventory transactions without any accuracy check such as a cycle count program or inventory accuracy audit. Most Class A processes are governed by the weekly performance review meeting where all process owners report progress and improvement of their process once a week, usually on Tuesday at 1 P.M. By having a management system in place, good process can continue through generations of employee turnover. More will follow on the topic of the Tuesday weekly performance review meeting.

Results

The component of Class A ERP that most people think of first is the metrics. Organizations often think they are ready for certification just because the metric performance is at Class A ERP levels of acceptability and are surprised by the shortcomings in process design and management systems. As you can see from the descriptions above, both of those areas of focus are important for high-performance organizations. Nonetheless, performance is the goal. Everybody can relate to measurements. We all have them in our lives, one place or another, and most of us have several. Class A ERP has numerous measurements, and to meet certification requirements, they need to be at Class A minimums. They also need to show sustainability. Review Table 2.4 for a helpful reminder of the Class A ERP thresholds of acceptability.

The measurements are only part of the picture, but obviously an important part. The certification audit generally not only looks for the correct level of performance, but also looks at actions and trends to see if the process is tracking a normal curve and shows ownership and sustainability.

SUMMARY

This audit process introduction should give you a sense for the profundity of the Class A ERP focus. At this point, after the first two chapters, you should also have an appreciation for the benefits of achieving it. While Class A ERP may not be the definition of world-class performance, it is impossible for any organization to achieve Class A ERP performance without being a high-performance company. At the same time, high-performance companies all have the Class A basics in place.

> While Class A ERP may not be the definition of world-class performance, it is impossible for any organization to achieve Class A ERP performance without being a high-performance company.

From this point on, the next several chapters will correspond exactly with the elements that make up the ERP business model (Figure 2.1), one at a time. Each chapter will deal with the objectives, implementation, required actions, metrics for performance, and management system elements for each of these elements. If you follow the steps as defined in this "how-to," your business is guaranteed to have predictable processes, and if you manufacture something somebody else wants to buy, you will even be guaranteed success. After all, as Pete Raymond (grandson of the late George Raymond, Sr., founder of The Raymond Corporation) said to me in a class one time, "This Class A is a good thing, but if we were making cannonballs, we could be Class A and still go out of business." He was right in some respects, but no improvement process — Class A ERP, lean, Six Sigma, supply chain management, total quality, Deming wheel, or any other — takes the place of business savvy or the understanding of business markets. Class A ERP just ensures that whatever it is you want to do, it gets done — accurately and on time. A valuable basic trait for any business!

3

BUSINESS PLANNING

The enterprise resource planning (ERP) business model starts with the top-management vision. In Class A ERP, the business plan is made up of at least three components: (1) strategic planning, (2) financial planning, and (3) business goals (Figure 3.1).

If we take business planning out of the ERP business model and look at it by itself in detail, it might be better depicted as a hierarchy of actions, only some of which are governed by the Class A ERP process itself.

Figure 3.2 depicts the elements in robust business planning and the processes that must exist for these objectives to take hold. The Class A ERP process itself does not dictate or prescribe methods for determining the vision of market success. Class A ERP performance *does ensure* the delivery of same. If a business expects a good-quality, sustainable experience within the performance objectives it has chosen, it just makes good sense that the business has processes in place that can deliver those results on a repeated basis. This means it must be easier to execute the process to the expected levels of performance than to not meet the goals.

There is an old saying: "Don't expect different results when you keep doing the same thing you did yesterday." There is a lot of truth to this. Too many

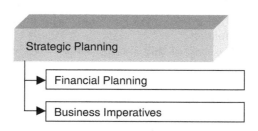

Figure 3.1. Top-Management Business Planning.

33

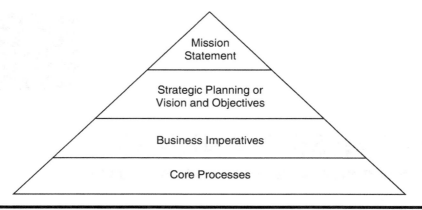

Figure 3.2.   The Elements in Business Planning.

businesses think improvement is about doing the same thing the same way, but working much harder and faster at it. Nothing could be farther from the truth. Sustainability means working smarter, not building a process that requires every mile in a thousand-mile journey to be a four-minute mile. It only works for a while because people get tired. Instead, we need an environment where people are challenged, but can also have fun. Core processes designed to meet business goals must have that important characteristic.

One tool designer and manufacturing business I was consulting with a few years ago was defining its objectives for the upcoming year. Ed, the general manager, had just finished documenting these goals. The first one was "increase sales revenue by 25 percent." (I see that objective a lot.) As a good consultant, I asked Ed where his confidence level was on that objective. He naturally stated that he had high confidence. Here is how the conversation went:

> *Sheldon:* "Did you have a similar objective this last year?"
> *Ed:* "Yes, we did. We had the same goal this past year."
> *Sheldon:* "And how did you do this past year?"
> *Ed:* "Well, this year was crazy. We introduced a new product line and the market pricing kept going down because of too much capacity in the market."
> *Sheldon:* "Was any of this a surprise?"
> *Ed:* "Not really."
> *Sheldon:* "Then you probably weren't really all that surprised that you didn't make the stated 25 percent increase in revenue, were you?"
> *Ed* (quietly): "I guess not."

Businesses often prefer stretch goals. In many applications, it is what I call World Federation Wrestling mentality — too often, it is mostly for show. In the case of revenue increases, the answer is much simpler than you might think. There are only two ways to grow revenue that I know of: (1) get old customers to buy more and (2) get new customers. Obviously, there are a multitude of actions a company can do to make either of those strategies happen, but the real key is in understanding what those actions will be. The first step is in asking the right questions. In Ed's case, the questions he needed to answer were: (1) What are the incremental products that existing customers are going to buy (and why) and/or (2) Who are the new customers (and why)?

If we answer the right questions concerning that set of objectives, we are forced into the thinking and strategies of what additional products or services we must offer. Beyond that thinking, we need to understand how. For example: How is pricing affected? Do I need to raise some prices and lower some others? What new products do I need to have designed? What is the competition going to do that will affect my performance and strategy? All good questions!

I am guessing that these are all strategy expectations you have thought of and you already know that you need to do them. You would be surprised to learn how many businesses I go into that have no real follow-through process for meeting their yearly goals. The good news is that Class A ERP process thinking is effective and simple. The required actions within Class A criteria are often what we would do anyway if we just thought about it a little longer. It is this simplicity that gives me my passion for the process. In the case of the people at the tool manufacturer, they found that when they asked the tough questions, they began to lose confidence in their 25 percent goal. This acknowledgment forced their hand and made them look more closely at competitive analysis. The result was to move up their current new product release date. This required some shifting of engineering staff, but made a big difference in their competitive position. They also initiated a business imperative review process that met monthly to follow up with project managers assigned to important projects critical to the business strategy. They had reviewed projects in the past, but did not have a formal and predictable review attended by all of the top staff members. This not only allowed for a more predictable objective meeting experience, but also orchestrated a much more fluid resource alignment process. They were much more effective at meeting their goals the following year.

Strategic plans should be reviewed and updated at least once a year. Class A predictability suggests that this review happen at the same time every year. This makes the meeting predictable, and process owners who are involved know exactly when this meeting will take place and can be best prepared for it. Normally, this meeting takes about three to four days depending on the size of the company. Smaller companies can finish this exercise in one to two days.

During the meeting, it is helpful to have each functional top manager give a presentation on divisional accomplishments and expectations. This information can be helpful for thinking globally. Attendees at this meeting are normally the top staff reporting to the CEO or president as well as a few key managers who report to them. If the meeting gets too large in number of participants, it can become inefficient. Probably ten to twelve participants is the upper limit. The danger is leaving some important input out of the equation. This can be remedied by having other key management or professionals not at the meeting present their input as a scheduled event for about one to two hours.

Additionally, there may be some value in getting an outsider to spur thought. Although this is not recommended in all cases, some clients have used facilitators effectively. The Raymond Corporation (the company I "grew up" in) would occasionally have an economist speak to the management team as the kickoff event for the strategic planning event. In our organization, it was a week-long planning process that always took place in August. The deliverables from this planning session were a strategic plan governing the organization for the three- to five-year period and a short list of corporate business imperatives.

BUSINESS IMPERATIVES

I have mentioned business imperatives a few times, but have not really stated in detail what they are, why they are helpful, or how these objectives fit into the Class A ERP picture. Every business has objectives. As already stated, some businesses are more serious about meeting them than others, but it is probably safe to say that all want this wish list to happen. When high-performance organizations establish the vision, they separate strategic goals from imperatives. Consider an example. In a business I worked with in the past, there was a strategy to diversify into additional markets. The business manufactured automatic and heavy truck components and had several patents on products that it believed would probably be affected negatively by replacement technology in the next ten to fifteen years. The company's strategic plan said its vision was to diversify into power generation, a distantly related technology from its existing auto parts designs. The strategic objective for that year was to "diversify into power-generating products." While this was measurable, for sure, it was clearly "visionary." There is a level below this in Class A ERP organizations called *imperatives*.

All intelligent managers can think of helpful things to do. In fact, the real problem with this intelligence is that they often think of *too many* things to do. Said a different way, many smart people in the business can generally name twenty things the company needs to do to improve competitive advantage, but it is the *real* leaders that can pick the five or ten goals a company will *really* do. By focusing the resource and prioritizing the goals in this manner, more actually gets done at the end of the day. Business imperatives are these "must do" objectives. When coaching top-management teams, I have always advised them to keep the business imperative list short. This is not easy to do. The linkage (see Figure 3.3) between the strategic and more tactical objectives as well as the core processes designed to ensure these goals are not only achieved but sustained is critical. This communication only happens though predictable management systems that, not so coincidently, are found commonly in a Class A ERP process.

TOP-LEVEL MANAGEMENT SYSTEMS

In Figure 3.3, you can see that Class A is not too engaged in the mission statement. In fact, the Class A ERP process is more about getting the objectives met than about determining the objectives. Having said that, to accomplish goals effectively, the goals must be realistic, and that is where the Class A ERP process helps in the development of objectives. Two important infrastructure events happen at least monthly within Class A businesses that require scrutiny on projects linked to the objectives. One is the sales and operations planning (S&OP) process and the other is the project review process. On a more global nature, Class A organizations conduct yearly strategic planning reviews to calibrate and adjust their vision, as was described earlier in this chapter.

These habits of follow-through force organizations to see the strength or weakness in plans even as they are being made. It is a thought process similar

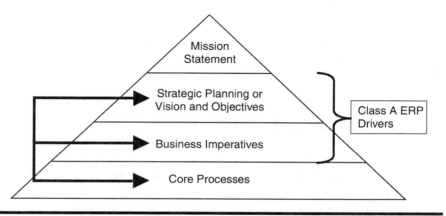

Figure 3.3. The Linkage Between Strategic and Tactical Objectives.

to "owning" something. Ownership makes you more cognizant of risks and potential gains. How companies choose the right goals is something not thought through in some organizations. There is no foolproof prioritization process, only business savvy and intelligent leaders. One method that has proven to be helpful in many businesses is documented in *The Balanced Scorecard* (Robert S. Kaplan, Harvard Business School Press, 1996).

TRICKS IN BUILDING THE RIGHT BUSINESS PLAN

The Balanced Scorecard is one of the best and yet simple approaches to business planning that I have found and used. In his book, Robert Kaplan suggests that managers look for opportunity and scrutinize four general focus areas:

- Financial
- Internal business process
- Learning and growth
- Customer

I like to add supply chain management and technology to the list, so the modified list would be:

- Financial
- Internal business process
- Learning and growth
- Customer
- Supply chain
- Technology

If you take Kaplan's advice and also apply Class A ERP management systems to this strategy development process, you will most certainly have a successful business planning model for your company.

FINANCIAL OBJECTIVES PLANNING

Few argue about the need to have good financial models, but many organizations do not practice the obvious. Family-owned businesses seem to be the worst, but surprisingly, even some publicly held businesses still have very poor financial planning models in place. The excuses are always the same: "You don't understand our business. The environment is always changing, and it is impossible to predict exact spending and revenue." Many of these businesses follow a plan that the *finance* people developed last October, for example. They then measure to this plan all year long. The accountants usually have their *divine vision* sometime around the 15th of October and come down from the mountain to communicate it to the rest of the organization, usually by Halloween.

The accountants usually have their *divine vision* sometime around the 15th of October and come down from the mountain to communicate it to the rest of the organization, usually by Halloween.

The reason businesses exist is to meet customer needs and grow equity for the stakeholders. Financial information is essential to measure both equity/stakeholder value and customer need fulfillment. In a Class A ERP environment, plans are always updated to the latest level of company knowledge. In the case of financial planning in a Class A ERP organization, there may often be two plans: (1) the original budget or operating plan submitted to the stakeholders prior to the fiscal calendar year start and then (2) a financial forecast that is updated at least monthly. The latter plan is the one that we will post measurements to and focus on in the Class A ERP management systems processes. Elements in the financial plan would include:

- Sales (in dollars or other monetary unit) by product family
- Cost per facility and/or division
- Cost per product family
- Inventory per product family
- Gross margin by product family
- Earnings before taxes (EBT)

These plans are to be twelve-month rolling plans linked to the sales and product plans. Obviously, each of these could have sublevel plans to review at the detail level or to report to management when metric performance is not to acceptable levels. For example, below the category of inventory by product family, there would normally be sorts by obsolete, excess, days on hand, etc. Since we are only talking about the top-management-level planning right at this point, in Class A ERP process, exact details can follow as resources are assigned and projects are scoped. Financial objectives do not stand alone. These objectives cannot be made without a full understanding of the sales and production plans as well as new product impact on market position and capacity issues that might exist. Many organizations have financial plans that are a result of taking the prior year's plan and adding 25 percent. (Sound familiar?) All of the financial planning areas mentioned earlier are preparatory for the S&OP process review detailed in Chapter 6.

INTERNAL BUSINESS PROCESS

The objectives connected to the internal business processes are along the same lines as what is often referred to as "core process" or "core competency." At The Raymond Corporation (manufacturer of material handling equipment), we knew that welding was a special skill we nurtured and grew internally. As the business grew and we looked at the options of expansion or subcontracting, welding always seemed to be a process we kept. The reason was simple. Raymond had some of the best welders within a very large radius (maybe the world!). Many of our products were highly engineered for specific applications and required some "hand" welding without fixtures or special tooling. Skill sets in the welding area gave us flexibility and high quality. Steel still played an important role in the design of forklift equipment and welding was going to probably be around for a while. All of these data aligned to say that we should probably continue to weld in-house, at least for the specialty equipment. I have not visited Raymond in a couple of years, but I would guess that it still does a lot of its own welding. It is a core competency — a process critical to its success and one at which it excels. A strategic objective in this type of environment could be "to develop the highest-quality welding process." On the other hand, a business imperative could be "to reduce process variation in the specialties welding area to a first-time quality of 99 percent." This type of goal would include several areas — engineering design, process engineering, and welding operations at a minimum. The difference between the strategic vision and a business imperative is the specificity, scope, and confidence. A rule that high-performance organizations follow in this regard is simple — every objective named as a business imperative *will be completed as planned in the next twelve months!*

> Every objective named as a business imperative *must be completed as planned in the next twelve months!*

This means that management must not only be on board with the ideas, but must also take responsibility to ensure that these imperatives are accomplished. These managers do not *do* the projects; rather, they ensure that the projects are completed successfully. This means supplying resource and capital, giving advice and counsel freely and frequently, and following up on agreements and promises — generally "carrying water" whenever necessary so that the objectives are met. Some managers are not willing to take that kind of risk, and it can be very risky in terms of personal commitment. Business imperatives are to be announced as required project completions. Only the best project managers are to be empowered on these imperative projects, and when support from engineering or maintenance is required, it is to be given. You probably get the picture by now. If the team has a need for a process engineer and one is not available, the top-management sponsor will step in and get the needed support. If there is a lack of good problem-solving experience, top management must recognize it quickly and get the help or training required. I will say it one more time: "business imperatives" are completed in the time frames that they are scheduled, normally within twelve months from assignment. I will avoid getting into best-practice project management at this point. It is a major linchpin with Six Sigma and a major integration point into Class A ERP process. Project management techniques will be covered in detail later when we deal with Six Sigma influences in Chapters 13 and 19.

One point that I *will* make here and will underline later is the importance of scope in accomplishing the business imperatives. Just because the business imperative may initially read like a single objective as it is listed in the business plan, in many more cases it is best implemented as a series of projects. It has been proven time and time again that if a twelve-month objective can be broken into several thirty- to forty-five-day projects, the overall productivity of the project completion will be increased. The following example supports the theory.

Class A ERP certification is often itself a goal designated as a twelve-month business imperative. In many of the organizations I have worked with, this is the case. The right way to implement this process is to have several smaller projects rather than just one huge company-wide project. For example, Class A performance in the inventory record accuracy space should not take more than 120 days and should have a specific team assigned to it. The Class A project team, as the top-level project in this hierarchy, may be made up of several team leaders of the smaller projects subservient to the Class A ERP certification business imperative. You can see where I am headed. As you might imagine, this approach is much more productive especially when there could be ten to twelve separate projects concurrently to support the overall objective. Inventory accuracy would have a specific process owner within the business that would report to the overall Class A implementation manager. The two managers will be focused in different areas, but share the common goal of Class A ERP certification. That is the wisdom and efficiency of dividing the process ownership and project components into pieces. Now back to the discussion on internal business processes.

The internal business process covers a lot of territory and is normally ripe with opportunity in most manufacturing organizations. The right way to analyze

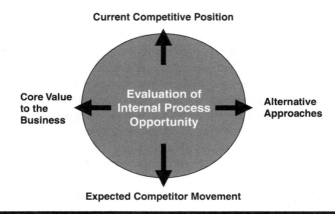

Figure 3.4. Evaluation of Internal Process.

opportunity is to list the core competencies and determine the competitive advantages that you already have and ones that could be gained with some insight and resource (Figure 3.4).

By asking the right questions and poking at all of the major processes in the business, high-performance organizations tend to create opportunity through creativity. These kinds of discussions spawn additional thinking — the kind that keeps you awake with excitement at 3 A.M.!

LEARNING AND GROWTH

Of all the topics, learning and growth opportunities is the biggest and yet often completely overlooked. Objectives in this area often can be the most powerful. This plays to the leaders who believe that people are their most important asset. In the consulting business, it is easy to understand. Without a network of skilled experts in specific fields, consulting companies would be businesses on paper only. It is true in all businesses at one level or another. You have probably heard the old adage that the best managers surround themselves with people who possess better skills than they themselves have. One manager, years ago, used to define good management effectiveness by a "delegation" yardstick. He used to say that strong, effective managers should be deceased in their office three days before anybody finds them.

Strong, effective managers should be deceased in their office three days before anybody finds them.

It is all true. Sustainable high-performance companies are never made up of less than effective or unskilled people. The best companies attract and retain the best people. GE is famous for it. It spends a large component of its annual budget on education and training. As a result, many of its top managers have gone on to run other large companies successfully. It is no coincidence.

In the mid-1990s, I was consulting with AlliedSignal (now Honeywell). The CEO at the time, Larry Bossidy (a protégée of Jack Welch, formerly with GE), measured managers in his organization by many areas he deemed important. One of those areas was hours of training and education. At AlliedSignal, there was a forty-hour minimum requirement for each employee, and each manager had to report his or her department's performance to the corporate office. The training was not arbitrary; it was always linked to the objectives of the department and ultimately the company. Skills assessments helped to target the right training for employees.

Prior to that, in the late 1980s, The Raymond Corporation also considered education to be a priority as it transformed from a business with declining margins to a world leader. Ross Colquhoun, the CEO during this transformation, held education and training as a high priority, and he would often brag that his department had the best ratio measured by "employees to tuition aid use." Interestingly enough, his definition of *best* was the *most* tuition aid used. This was at a time when Raymond was reorganizing and cutting costs in every area possible. It is arguable how much of the company transformation was due to this investment, but most who were involved in it would probably agree that it had some positive impact. The payback is always there as education and training are linked to business need. The exercise of using this topic to determine possible business imperatives is a very satisfying one when done right. It is an opportunity to invest in human capital and improve both the business and its members.

Looking at the skill set gap from the advantage of a global viewpoint that takes into account anticipated market shifts and technology changes (both areas that play a role in an organization's competitive advantage) makes the task of choosing the final list of business imperatives easier, or if not easier, it at least assures more confidence in the final list's appropriateness. After all, this is the list for final resource assignment, and in most organizations, resources are precious. The questions are always the same in manufacturing businesses. The answers are not always consistent, however. This again is where business savvy and insight pay off, as well as getting homework done prior to the business imperative assignment.

Brainstorm the possibilities using the topic focus from Figure 3.5. After scrutiny, compare the results with survivor ideas from the other category groups.

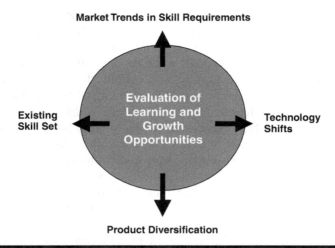

Figure 3.5.   Evaluations of Learning and Growth Opportunities.

Do not weight results for any particular topic until the final group of possibilities has been determined.

CUSTOMER INFLUENCES ON BUSINESS IMPERATIVE CHOICES

Let's face it, customers define success. Customers are a very influential component for use in the prioritization of *any* resource assignment list. It also is a difficult topic to guess correctly. That is why it is important to think about affecting customer behavior and not just trying to guess what customers are going to do. It ultimately comes down to consumer decisions, and we all know how difficult they are to predict. Even capital goods suppliers that sell to manufacturers and not directly to consumers are ultimately affected by consumer requirements. Fickle consumers (us) are never completely predictable even in efforts to change their behaviors (Figure 3.6).

Since adding new customers is usually a good thing, it may be one of the first areas to cover. What would we have to change to attract twenty-five million new customers? What is it that attracts our existing customers and why does it not attract noncustomers? Are they different in some way? All of these questions are the route to robust goal planning. This can even become easier if the vision is clear in terms of where the business needs to go. Then the new customer question has parameters as you start the process. This imperative driver is always fun and provocative.

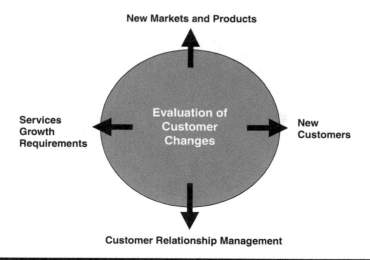

Figure 3.6.   Evaluation of Customer Changes.

SUPPLY CHAIN INFLUENCES ON BUSINESS IMPERATIVE CHOICES

In today's environment, the topic of supply chain can be a frustrating one. There are so many good choices on the surface that turn out to be bad ones once implemented as there are many factors that play into the decisions. We will get into this topic specifically in Chapter 10, but it is worth mentioning some of the pitfalls here. Many factors can create havoc overnight, such as the price of oil or transportation routes involved in disputes, fluctuation of monetary values, or more recently, terrorist threats in a particular region. A manufacturing organization has little influence to stop or avoid any of these. Nonetheless, great decisions have been made that resulted in low-cost producers being established on the other side of the globe. Many of these decisions have worked well. Here comes that business savvy requirement again. Little risk often adds up to little gain.

Keep in mind that none of these topic guides is the "end-all." Instead, each is designed to get the right topics on the table and help with the process of determining the right business imperatives. As you can imagine, this same process would work to help with the vision development as well.

When it comes to critical component suppliers, the process should be thought of as the lifeline, not the supply line.

When it comes to critical component suppliers, the process should be

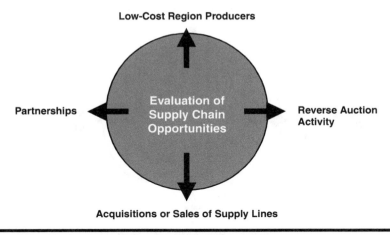

Figure 3.7.   Evaluations of Supply Chain Opportunities.

thought of as the lifeline, not the supply line. If you were going underwater in a diving suit that required someone to be on the surface checking the air supply piped to you, wouldn't you want the best of the best? Your survival depends on it. It disturbs me when so little thought is put behind the risk when moving critical components offshore. Building relationships or even partnerships often allows the best competitive advantage (Figure 3.7).

TECHNOLOGY AFFECTING BUSINESS IMPERATIVE DECISIONS

In the forklift market that I grew up in, it was often a market-leading activity to be the first to apply automotive technology to forklift trucks. Things like self-diagnostics and digital technology were well understood by the marketplace before they found their way into a lift truck. Some markets, such as consumer electronics, are not so fortunate. In any case, technology is a major factor in determining the next steps your business needs to take for competitive advantage.

In the mid-1990s, I did some consulting work with a nail manufacturer in the Midwest. This was a unique experience as the equipment and technology at the time were about as old as any I had ever seen. The company was using machines daily that had been in the factory for over seventy-five years! But interestingly enough, even those nail folks found themselves fighting technology migration. You cannot avoid it! In fact, the technology for the nail manu-

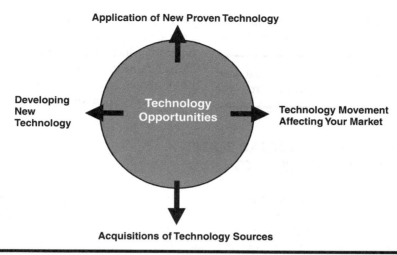

Figure 3.8.   Technology Opportunities.

facturing industry is changing every day, and quality and process variation become a major factor when automatic equipment is applied.

When asking the questions about technology, it is a good idea to bring creative people into the discussion. Every company has them — the dreamers. Engineers are often ahead of the curve, at least the best ones are. I also like to reread leading-edge books on the topic (Figure 3.8).

At this point, we have discussed several topics that will get your team thinking about opportunities for the final short list of business imperatives. Once the list is assembled, use normal decision-making processes to limit it to the right number.

HOW MANY BUSINESS IMPERATIVES SHOULD YOU END UP WITH?

There is no perfect formula for the list length when it comes to business imperatives. Years ago I went to "Jonah" school. A Jonah is acknowledged as an expert in Eli Goldratt's (author and scholar) Theory of Constraints. In that session, the lesson was that if you manage the orifice in the flow, or the smallest point in the process, the output would be predictable. This process is applicable in determining the correct number of business imperatives as well. We all understand that if a work center is overscheduled, the result is less throughput. If we overcommit our project resources, we create the same problem. The original thought was maybe the most important, which stated that the business imperatives must be completed in the next twelve months. That alone is the biggest influence in determining the correct number of top-priority goals. Empower no more than you are willing to fund and assign resource to.

> Empower no more than you are willing to fund and assign resource to.

LINKING THE FINANCIAL PLANS WITH THE S&OP PROCESS

The essence of the Class A ERP process is to define linkages within the business planning process. Through years of evolving the ERP process criteria, companies have determined that there is an optimum level for detail planning at the top-management level. This detail generically is referred to as the "product family" level. If product family has a specific meaning to you in your environment, do not get too hung up trying to apply your definition. It can be different in every business, as product family does not necessarily have a common definition.

As you can see from Figure 3.9, there is an optimum level of detail for the product family designation. This needs experimentation. It is difficult at best to try to define a generic formula for pinpointing the correct configuration that would work in all manufacturing companies. It just is not that simple. It is common for a company to change the product families several times before they settle down to consistent groupings. The discussion on the financials has to start

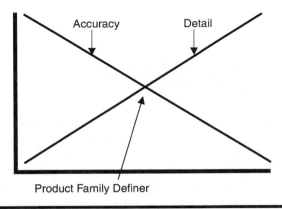

Figure 3.9. Accuracy Versus Detail Comparison.

here, as it is critical that the groupings in the financial planning match the demand plans and the supply plans exactly. It is this insistence that allows the real power of the S&OP process to work.

PRODUCT PLANNING IN CLASS A ERP

Product planning is a key to successful market competitiveness and share growth. While this book is not intended to be about product planning, there are a few focus areas worth including in your Class A ERP understanding. You will see this importance as we move into the next business model process, demand planning. Good product planning in a business establishes market drivers and assumptions that the rest of the business can build process around. This would include answers to the following questions: What markets are we to serve? How are these markets defined? What market position do we expect to be in? Answers to these questions are especially important to businesses that want to produce cost-effective products and capacity in anticipation of the demand. In many companies, anticipating demand is necessary due to lead-time requirements of capital improvements. Too many times, these requirements can be out of sync with customer lead times once demand has been introduced into the market. It is not unusual for business imperatives to develop from product family analysis. For example, the strategic plan in one business is to diversify into a related product offering. Timing may be critical for successful market impact. The product specifications have been developed and capacity has been procured for the first product. A business imperative may be to have the first production units shipped by a specific date. All the elements of Class A ERP begin to play into the success or failure of the new initiative.

THE CLASS A MEASUREMENT FOR BUSINESS PLANNING

There are some assumptions made as the business planning process, which is quite complex, comes to the point of measuring results. The main assumption is that profitability is a direct result of good business planning. This keeps the metric simple and also reminds everybody what we are really here for: to make money! The Class A ERP measurement requirement is 95 percent accuracy of the financial planning process. Keep in mind that the Class A accuracy may not be as high as the CEO of your organization requires; we are only talking about a threshold of acceptability for Class A criteria. It is important to understand that fact because often when the profit plan calls for $10 million, $9.5

million is not acceptable to top management, and that is not only reasonable but understandable.

In all of the other metrics within Class A ERP criteria, plan accuracy is the goal, meaning that deviations from plan to the *positive* are looked at as process variation just the same as variation to the negative. In Class A business planning, we do not look at high profitability as bad. When a plan of $10 million results in an actual profit of $10.7 million, the measurement is still 100 percent with no performance penalty for beating plan. As you will see in the rest of the metrics within Class A ERP, this is not true in any other measurements. (By the way, notice that the performance in that example was not 107 percent!)

Business Plan Metric Formula for Profit Centers

$$\frac{\text{Actual profit results}}{\text{Planned profit results}} = \text{Percent of accuracy*}$$

In many organizations, Class A ERP is applied separately in all facilities and not just once for the whole company. In most of these organizations, the facilities are not necessarily all profit centers. In facilities where the financial obligations are treated as cost plans rather than profit plans, the metric is very similar.

Business Plan Metric Formula for Cost Centers

$$\frac{\text{Actual cost results}}{\text{Planned budget}} = \text{Percent of accuracy**}$$

Be aware that, in either a profit center or cost center, if the plans are sandbagged and the results are continuously beat through higher profits or lower costs, the "better than plan measurement forgiveness" should be abandoned. If an organization beats its plan for three out of four months, it is most likely buffering its plan unnecessarily and should be penalized for it in the metric. In these cases, the metric would go to an absolute percentage of deviation, as do the other metrics in Class A ERP. In one family-owned pack-

---

* If actual profit is higher than the plan, the measure is automatically 100 percent. No performance above 100 percent is allowable, since it is really a case of bad planning if the number is way off target.

** If actual costs are lower than the plan, the measure is automatically 100 percent. No performance above 100 percent is allowable, since it is really a case of bad planning if the number is way off target.

aging business, the profit plan was beaten month after month. Usually the differential was significant. After about six months of this, it was apparent that there was a lack of commitment on the part of the management team. Class A requires two aspects of accuracy: a good plan and good execution. When plans are continuously beaten, it normally means there has been reasonably good execution, but it is also as likely that there is bad planning.

> When plans are continuously beaten, it normally means there has been reasonably good execution, but it is also as likely that there is bad planning.

TALENT REVIEW

What could possibly be more important to any business than its people? In high-performance businesses today, there is an acceptance that people are the most important resource. This is demonstrated by investments in hiring the right people, in growing existing people, and helping nonperforming or subpar employees improve or get reassigned to duties more suited to their talent. To keep the movement in a positive direction, Class A organizations evaluate talent each year to make sure there is an acknowledgment and documentation on who and where the best talent is in the business. Many refer to these as the "high pots," the slang abbreviation for "high-potential employees." Once a year, each manager in the business rates the associates in his or her respective department for promotability, motivation, and special skills. Many organizations also use this exercise to do succession planning. This entails developing a backup list of people who could easily or effectively replace existing key positions should the current person leave or get promoted.

> Having high-potential employees in your department is essential in creating a continuously improving environment.

Having high-potential employees in your department is essential in creating a continuously improving environment. Normally, it is these people who ask the tough questions and challenge status quo. Contrast this with the "cash cow" or longer-term process participants. Cash cow employees are highly knowledgeable resources who understand the existing process completely and often have many years of experience. This gives them the advantage of having seen numerous exceptions to the process happen over time. Good departments have a healthy mix of both "cash cows" and "high pots." Without going through the evaluation, it is difficult to be cognizant of the opportunities in this area.

PROCESS OWNERSHIP OF THE BUSINESS PLAN

In the business planning space, there is no doubt about who the process owner is. In a multifacility Class A ERP environment where each facility is being certified separately, it is the top manager in each facility who owns business planning for that facility. In a plant, it would normally be the plant manager. In a division, it would be the general manager or VP. In the corporate head-quarters, it is the CEO. The process owner of any process in Class A ERP literally is expected to own the process and the performance. If the Class A minimum performance is not being met, it is up to the process owner to either fix it or make sure it gets fixed. This includes reporting to management system events. In the case of the business plan, as a minimum, the management system events are the yearly strategic planning review, the yearly talent review, a monthly project review, and the monthly S&OP process.

SUMMARY

In a Class A business plan, there will be several process elements in place that establish both solid objectives as well as financial results. The main focus areas for Class A ERP in Figure 3.10 are below the mission statement. Class A is more about making sure the right things happen than establishing the mission of the business. Once management has established the vision, it is easy to get excited about making it happen. This is done through metrics and the management system. The list would include:

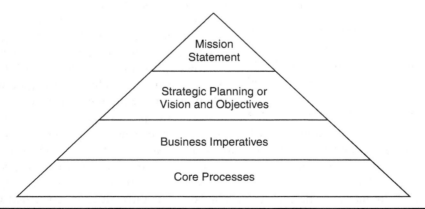

Figure 3.10. Business Planning.

- Strategic planning — three- to five-year horizon
- Business imperatives — defined and communicated
- Financial planning process (budget) — both yearly budget as well as a monthly, updated, twelve-month rolling plan by product family to support the S&OP process
- Financial plan accuracy metric — posted monthly

The business planning process sets the business up properly. The next step in the Class A ERP process model is demand planning.

4

SALES AND MARKETING PLANNING: CREATING THE DEMAND PLAN

Probably no single process within the enterprise resource planning (ERP) business model gets more attention than demand planning. I spend a lot of time with operations people, and demand planning is a popular topic with most of them. The popularity comes from the frequently observed wide-band process variation introduced into almost every business by its customers. If your business does not have unpredictable customers, then you are the exception! In many businesses, the demand plan or forecast is actually done by the operations people because the demand-side folks have not willingly come to the planning table yet to give forecasting a decent chance. In a Class A ERP environment, it is necessary that there is a solid handshake between the demand process and the operations side of the business. This handshake culminates in the sales and operations planning (S&OP) process. Not only is this partnership attitude important for minimizing risk, it also creates an environment where everybody has shared goals and understands how variation from their part of the business affects other parts of the operation.

The inputs are what make the quality of the demand plan (Figure 4.1). If the emphasis is on the right tasks and they are linked to the business plan properly, the results are fun indeed! For this to happen, there have to be well-communicated, well-understood objectives with shared ownership in the outcomes. Let's look at each input individually.

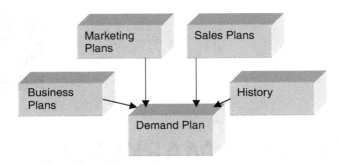

Figure 4.1.   Class A ERP Demand Plan Inputs.

BUSINESS PLANNING AS AN INPUT TO DEMAND PLANNING

In businesses that are continually developing competitive advantage, it is important to have the plans and actions tightly linked. As management direction influences new product development, new markets, and service offerings, it is extremely helpful if everyone is on the same page. The business plan inputs (Figure 4.2) are probably the most important to the company's success. It is these inputs that direct specific resource alignment in the marketing and sales area and redirect goals. It is not about following the customers; it is about leading them when at all possible.

If management has prioritized a new service, actions need to result that affect this outcome. Metrics also need to be implemented to gauge that the results are as predicted. A company I work with is a first-tier supplier to the food industry. The company decided to focus on just-in-time deliveries for competitive advantage with its best customers. The corporate business plan

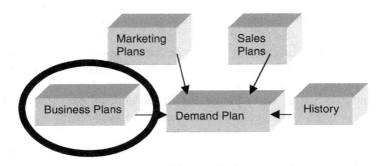

Figure 4.2.   Class A ERP Demand Planning Inputs: Business Plans.

established the objective of building new plants across the street from the company's best customers, but not until there was a partnership agreement from these customers. Longer-term agreements, in this case, mean less inventory requirements for the customer to hold and shorter lead time. In the short term, it requires the sales force to sell the idea and get the signature. This may not have been the sales plan without the business plan input. There are many more examples of business plan input requiring a shift in resource and effort. Some might include a new service offering to customers that only buy product now, new markets being entered, selling into Asia if a new experience, raising prices in a tight market for strategic purposes, etc.

MARKETING PLANS AS AN INPUT TO DEMAND PLANNING

When operations people think about forecasts, they often think about *guessing* what customers are going to do in terms of product demand. It is common to hear statements like, "Man, if we could just know ahead of time what customers were going to want, we could run very effectively and with very little inventory." This kind of thinking is often misdirected. In high-performance businesses, the correct thought process for demand planning is not to *guess* what customers are going to do, but instead *affect* what they are going to do (Figure 4.3). If you think about it, this strategy makes much more sense when you link to the business planning strategy. Most businesses today are not

This means that marketing must estimate what impact each of its strategies will have on customer behavior. That requires commitment from the marketing folks. In Class A ERP, it is referred to as process ownership.

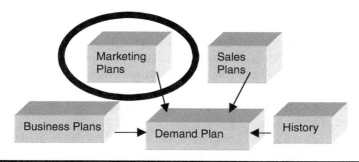

Figure 4.3.   Class A ERP Demand Planning Inputs: Marketing Plans.

trying to repeat yesterday; they are trying to do things differently to improve their position in the marketplace continuously. This means that marketing must estimate what impact each of its strategies will have on customer behavior. That requires commitment from the marketing folks. In Class A ERP, it is referred to as process ownership.

For example, let's say that the business plan has outlined two new products for design and introduction. One product is in a totally new market, and the other is a new design in an existing market. Marketing will usually align with the new product introduction teams and have its people assigned to each focus area. If the new product is also in a new market for this example company and there are three marketing ideas to implement, marketing needs to link its plan to some sort of expectation. As an old boss of mine used to say about *everything*, "There's always math behind it." We need to get to the math. Let's say that the three ideas include one trade show, one customer event, and one advertising campaign. If the marketing group puts a return value (increased demand) on each, some effect would be calculable on expected demand. The assumption, in this example, is an estimate of 30,000 units of demand if the company did nothing except introduce the new product through existing markets. Splash from the planned trade show might add an incremental 10,000 units. The estimate is best done if the potential new customers are listed individually with anticipated "buy" quantities. The customer event designed to create excitement for a few of the biggest potential customers for this product could add another 50,000 units, and lastly, the advertising campaign might have an effect of 25,000 incremental units. Doing the math, this marketing team could add 30,000 (pre-existing potential) plus 10,000 (trade show), plus 50,000 (customer event), plus 25,000 (advertising campaign), for a total of 115,000. If you are thinking that it would be a miracle if the 115,000 was accurate, you would probably be right. Marketing needs to estimate process variation just like operations people have to. In many businesses, it is better to have the forecast a little off on the low side on incremental units than to have way too much inventory and nobody buying it. Marketing needs to factor the estimates.

What makes this process powerful is the process ownership. The factoring has some risks, and the marketing people need to know that they will be measured on their accuracy. High-performance organizations should expect 85 to 90 percent accuracy overall in demand forecasting once accountability and ownership have been established. Remember that, as discussed in Chapter 3, this accuracy is measured at the product family level, not stockkeeping unit (SKU) level (Figure 4.4).

For the product family, companies see the highest level of detail at the highest level of accuracy. This optimum level of detail takes some experiment-

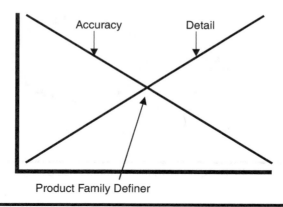

Figure 4.4. Accuracy Versus Detail Comparison (from Chapter 3).

ing. If both the marketing and the operations people have the shared goal of accuracy in this metric, the job can actually be fun. The following rules work best. Writing them on the wall in the main conference room could be helpful. If organizations can get this agreement between the two groups, mountains can be moved.

1. The operations people need to stop pointing fingers at the marketing people for missing forecasts so badly and help them succeed. Forecasting accuracy is much more difficult than scheduling products accurately.
2. The marketing people need to step up to the bar and understand and acknowledge that the farther the forecast is off, the more cost potential is generated for the company products.

SALES PLANNING AS AN INPUT TO DEMAND PLANNING

The next input area to look at is the sales plan (Figure 4.5). This plan is more calculable than the marketing plans. In many businesses, there are experiences to work from in terms of closure rates and sales cycle times. In capital equipment, for example, there is normally a relatively long sales cycle, but it is known and can be reasonably predictable over a broad customer base. Other consumer products can have no time at all. In that case, the customer reaction is also relatively predictable. Sales are often linked to seasonality or some sort of cyclicality. One clothing manufacturer I worked with specialized in dance costumes. Because of the seasonality of its product, 80 percent of its sales were

Figure 4.5. Class A ERP Demand Planning Inputs: Sales Plans.

seen in just three calendar weeks of the year (dance recitals are always in the May time frame). Another client, an ice cream cone bakery, did the lion's share of its business in the four months of the summer. Still another client, a kitchen appliance manufacturer, found its sales very seasonal as well. The most familiar saying I hear in businesses is, "You don't understand, our business is different." The truth is, there are more similarities than differences in most businesses.

> The most familiar saying I hear in businesses is, "You don't understand, our business is different." The truth is, there are more similarities than differences in most businesses.

In the 1980s at The Raymond Corporation, we missed plans too frequently. At one point in the early 1990s and in the early days of Class A, we started to calculate sales and the "math behind the numbers." During this exercise, we discovered that our plan of increased sales and profit was probably not possible. It was calculated that with the number of people presently in the sales role and the projected closure rate, given the number of expected market opportunities, the math showed we would fall severely short of plan. The conclusion was that we did not have enough "feet on the street." If we had not taken a statistical approach to this, we would have taken much longer to come to the correct conclusion. It even might have been too late. The story has a happy ending, however. The salespeople were added and Raymond became a market leader once again and still is today. Justification to add salespeople required the math (facts) behind the process. Understanding the root cause happened because we had everybody at the table and involved.

Getting the math behind the numbers in the sales function is critical. This can mean predictability in numbers of customer calls, frequency of visits, follow-up on late orders, number of cold calls, etc. Again, the business plan is a major driver in the measurements that could be appropriate in your business, but the sales plan is close behind.

HISTORICAL INPUTS TO THE DEMAND PLANNING PROCESS

The last input to the demand planning process is the one that is usually thought of first, historical data (Figure 4.6). The reason it is thought of first is most likely because, in many businesses, it has been the role of operations to do forecasting. Since the operations people know very little about what is happening in the marketplace because they are not out in it, the only helpful data they have is history. In many businesses, history projected forward is interestingly accurate. That is not always good, however. Businesses that can accurately predict the future using the past are often not changing the landscape in what for many are changing markets. This can often come back to haunt the organization in lost share. This is not to say that history is not valuable. Quite the contrary, it is valuable. Within the data are the seasonal stories and the normal cyclicality that are so necessary to understand. It is just not the entire story, though, in high-performance businesses that are executing strategies to increase market share.

Today, there are also many good software tools, such as Futurion®, Demand Solutions®, and many others, that can organize data and historical information to help predict future events statistically. They are helpful in organizing customer data and applying the data to the master production schedule. These tools have come a long way in the last few years and should be considered if your ERP business system does not handle these activities proficiently.

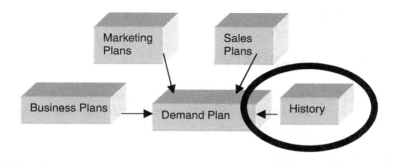

Figure 4.6. Class A ERP Demand Planning Inputs: History.

The message to end the discussion on history as an input to the demand plan is this: History is the most difficult input to use accurately for the demand planning in businesses that are making waves in their markets. If you are shaking things up with new products and services, history becomes less likely to be accurate or even valuable. Historical data can be manipulated in many ways as well. Software forecasting equations that allow weighting in any period you wish actually allow you to get whatever result you want. That is good and bad. Keeping history in perspective is helpful. Use it wisely!

UNDERSTANDING THE OUTPUTS OF DEMAND PLANNING

The demand planning process is done to help the business understand profit potential. Indirectly, it sets the stage for capacity, financing, and stakeholder confidence. The forecast is probably the most valuable input to the risk management of the business. At the S&OP process meeting, top management will evaluate the likelihood of the demand plan being executed accurately and will commit resources and cash accordingly. Demand planning is not a one-way communication or plan that gets thrown over the wall to manufacturing. Instead, it is a handshake agreement between all the parties. In essence, everyone tries to keep everyone else honest. This system of checks and balances helps with the risk management process.

> Demand planning is not a one-way communication or plan that gets thrown over the wall to manufacturing. Instead, it is a handshake agreement between all the parties.

The output of the demand plan is the forecast (Figure 4.7). It should be divided into product families and have a horizon of twelve rolling months. This is to meet the requirements of the S&OP process (more details on the S&OP process are given in Chapter 6). Table 4.1 shows a typical demand plan.

THE WEEKLY DEMAND REVIEW

The demand plan is a moving document due to the simple fact that customers are generally ill behaved. Each week, we are smarter than we were a week ago, and it is foolish to ignore new data. For that reason, each Friday there should be a demand review held with all product managers, plant managers, and master schedulers. In most businesses, this allows changes in the current week to be implemented quickly and baked into the schedule changes for the upcoming week. In some businesses, this may mean changes to the supplier signals (make

Figure 4.7. Class A ERP Demand Planning Outputs.

to order, engineer to order), in others changes in the manufacturing schedule (make to order, assemble to order), and in others changes in distribution (make to stock). It just depends on your business's inventory strategy — whether you keep it in finished goods, in components, or at the supplier's facility.

Table 4.1. Date: 2 February 20___

Product Family	Jan*	Feb	Mar	Apr	May	Jun	Jul	Aug	Sep	Oct	Nov	Dec	Jan
All Numbers in Thousands													
Line one	30	30	20	25	34	35	45	78	67	50	35	35	30
Actual	30												
Line two	30	27	22	21	20	25	15	20	19	16	15	12	10
Actual	29												
Line three existing	60	80	86	90	88	89	90	78	65	50	35	30	32
Line three new**								20	25	50	75	80	90
Actual	66												
Line four	50	51	55	65	64	66	73	78	67	55	56	55	50
Actual	54												
Line five	5	8	6	7	7	7	8	9	10	9	7	6	6
Actual	5												

* The January column in this example has both planned and actual numbers from the month's performance; the balance of the months are planned quantities.

** New products are generally reviewed separately from existing products due to the high-risk opportunity.

Table 4.2   Monthly Demand Review Schedule

	Mon	Tue	Wed	Thu	Fri
Week 1		S&OP			Demand review
Week 2					Demand review
Week 3					Demand review
Week 4					Pre-S&OP review

Sample month where the first day of the month happens to be a Monday.

The demand review should happen at the same time each week with the same players. A standard format should also be followed to ensure the right things are discussed:

1. Review current week's accuracy of forecast for each product family
2. Review upcoming week's forecast and any changes necessary
3. Agree on any required adjustments to the production plan
4. Determine any effect on the month's revenue or profit
5. Review actions and agreements left from last week and new ones for this week

Using Lotus Notes®, Windows Outlook®, or other scheduling software on your server, schedule the meeting as an ongoing, reoccurring meeting (Table 4.2). This way, it automatically shows up on everyone's calendar who should attend and no one can say they did not get the e-mail or notice of the meeting. The meeting can be in a central meeting room or it can be by phone. Conference calls are often the only option for larger companies where logistics are impossible or too costly to meet in a common location.

At one roofing company with headquarters and sales in Massachusetts and manufacturing in North Carolina, the weekly demand review is done with a video conferencing setup each Friday. Each week, management from the factory meets "face to face" via camera and television with management from the demand side hundreds of miles away. It works well, and the addition of video allows improved communication, not always evident in phone-only remote conversations.

The attendees to the weekly demand review can be somewhat different from one business to another due to size and structure as well as job titles, but the general expectation requires process owners from the demand plan to be on the call or at the meeting. A normal list of attendees might look like the following:

1. Product managers (demand side)
2. Plant managers (supply side)
3. Operations VP (supply side)

4. Master scheduler(s) (supply side)
5. Optional attendees:
   a. Sales VP (demand side)
   b. Marketing VP (demand side)

In some larger organizations, people come in and out of the meeting according to prescheduled time slots for specific product families. For example, the product manager for Line 2 does not need to be on the call when Line 1 is being discussed unless it is of some specific interest due to shared resources, etc. If multiple plants are involved, usually master schedulers at each facility are only interested in lines that affect their facility and schedule. This discussion is not to keep people out of the meeting for any other reason than to minimize their time commitment. Anybody who wishes to and has time is welcome to listen in on any of the meeting. The weekly demand review meeting is a great communication tool between the demand and operations sides of the business.

When someone is unable to make the call due to illness, vacation, or customer requirements, a replacement is named and attends with full decision-making authority in the person's absence. Once a month, usually the last meeting in the month, the agenda changes in preparation for the soon-to-happen S&OP process review. This meeting, just prior to the S&OP, is the monthly demand review where the monthly results will be reviewed to assess the total deviation from the original monthly plan as it was locked for review at the beginning of the month. The Oli Wight consulting organization's methodology often refers to this meeting as the pre-S&OP process. Because of Oli Wight's influence in this space, it is probably appropriate to call it that. Regardless of what you call this meeting, the weekly demand review prior to the S&OP needs to have the monthly emphasis not necessarily part of the normal weekly meeting.

The agenda for the pre-S&OP meeting is not so unlike the weekly demand review. The biggest difference is twofold: (1) the discussion around demand and production plan variation is more of a summary for the month rather than just the last week's variation and (2) there is a full acknowledgment that the S&OP meeting is only a few days away and questions need to have both answers and proposals for improvement. This acknowledgment makes the meeting more productive knowing that there is a deadline pending. With everybody in the room to make important decisions, the process can be quite efficient.

PROCESS OWNERSHIP IN THE DEMAND PLAN

The process ownership in the demand planning space can differ from business to business, again because of differences in organization structure and job titles.

At The Raymond Corporation, the VP of marketing had all the product managers reporting to her. Product managers had a huge influence on the business and were aligned by product family — the same divisors that are used in the S&OP process and demand planning. Product managers determined marketing plans and helped manage the sales force policy at the distributors, which were independent dealerships. The product managers were responsible for gathering information from the dealers, massaging it to their liking, and delivering forecasts to the VP of marketing. In turn, she would deliver these estimates to the S&OP process, where the records would be reviewed and blessed accordingly. She was ultimately the process owner for demand planning, although she got a lot of help from the VP of sales, and both were obviously present for the demand accuracy review during both the end-of-month demand review (pre-S&OP) and the S&OP.

I have worked with other smaller businesses where the VP of sales is also responsible for marketing. In these businesses, everybody from the demand side of the business reports to the VP of sales. There is little question who the process owner is in these organizations — the VP of sales. One approach to avoid, if possible, is to have the process ownership for demand planning accuracy delegated too far down the organization. It is the president or CEO's job to ask the tough questions. The VP level is normally the appropriate level for process ownership to reside for the demand plan accuracy, and this person answers to that accountability. It takes a lot of help from the organization to get it right, and the more handshakes there are, the more likely there will be a successful process.

Process ownership means that the "buck stops here." Accuracy is the goal, and the process owner is responsible for providing evidence of both learning and actions to improve the existing demand accuracy by product family. This is a role played out at several management system events throughout the month, the weekly demand review, end-of-month demand review, and the S&OP meetings.

THE MEASUREMENT PROCESS FOR THE CLASS A DEMAND PLAN

The measurement process for demand planning accuracy is simple. The calculation is always done from the product-family-level accuracy. The normal acceptability threshold is 90+ percent. The posted measurement is the average accuracy by product family. Some people get a little uncomfortable with average accuracy because accuracy itself is depicted as an average. Table 4.3 shows how demand accuracy is calculated for Class A certification. The data

Table 4.3. Demand Plan Accuracy Measure

	Accuracy
Product family 1	100
Product family 2	97
Product family 3	90
Product family 4	92
Product family 5	100

Total performance = (100 + 97 + 90 + 92 + 100)/5 families or 95.8 percent

are from the January performance shown in Table 4.1. The accuracy is determined from the accuracy of each family. The family performance numbers in January from Table 4.1 would calculate per the example in Table 4.3 (see also Table 4.4).

One of the early questions about Class A measurements in the demand planning space is about how to interpret the measurement rules. The forecast

**Table 4.4. Demand Plan Accuracy Overall Performance
(Date: 2 February 20___)**

Product Family	Jan*	Feb	Mar	Apr	May	Jun	Jul	Aug	Sep	Oct	Nov	Dec	Jan
All Numbers in Thousands													
Line one	30	30	20	25	34	35	45	78	67	50	35	35	30
Actual	30												
Performance	**100%**												
Line two	30	27	22	21	20	25	15	20	19	16	15	12	10
Actual	29												
Performance	**97%**												
Line three existing	60	80	86	90	88	89	90	78	65	50	35	30	32
Line three new**								20	25	50	75	80	90
Actual	66												
Performance	**90%**												
Line four	50	51	55	65	64	66	73	78	67	55	56	55	50
Actual	54												
Performance	**92%**												
Line five	5	8	6	7	7	7	8	9	10	9	7	6	6
Actual	5												
Performance	**100%**												
Perf. total	**95.8%**												

* The January column in this example has both planned and actual numbers from the month's performance; the balance of the months are planned quantities.
** New products are generally reviewed separately from existing products due to the high-risk opportunity.

and the actual are not always what they seem. For example, most companies have blanket order agreements with their best customers. This makes good sense. It eliminates unnecessary documentation and paperwork flow. Many companies, however, make the mistake of using the blanket order as the "actual order" in the demand planning metric. This often does *not* make sense because the customers normally give a blanket order and then later change the quantities or schedule just prior to the ship date. Most of the time, blanket orders are just forecasts. No argument from the point that these signals are pretty firm. Most supply chain agreements include obligations for product on blanket up to certain limits. Nonetheless, however, these signals are really not much more than a "heads up" that the customer is about to schedule something and you better be ready with some anticipated inventory or capacity.

The metric is designed to measure accuracy of the forecast. Unless you have a special understanding with your customers that makes these blanket orders firm, acknowledge the blankets as forecasts only. The real order is the confirmed schedule a week or so out from shipment. At one automotive supplier I worked with in Europe, its customer, Volkswagen, would release blanket purchase orders several weeks into the future and also communicated releases to these blanket orders daily. Each week, both the blanket and the releases from the blanket were updated for the metric. The sales organization in the supplier company was responsible for forecasting how many firm, released units per product family the manufacturer would get scheduled by Volkswagen for the month, regardless of the blanket order quantities received. The closer you get to the end-item consumer in the food chain, the more applicable this rule will be. The reason is simple — unruly customers. It is preparation for the unruliness that makes good suppliers and happy customers.

Demand planning is a monthly metric. The demand plan is updated as required throughout the month, obviously, but for measurement purposes only, the "plan of record" remains locked. This plan of record is renewed each month. This forecast is locked at the S&OP each month and is measured for monthly accuracy. Reporting is only normally done in percentage format at the "end-of-the-month demand review" (pre-S&OP) and the S&OP meeting. The focus should not only be on accuracy, but also what can be learned from the inaccuracies experienced. Demand planning in a Class A ERP organization is about shared goals, process ownership clearly defined, a management system called the S&OP, and the weekly demand review to keep everyone communicating properly. There will be more about the top-management planning processes as we move into Chapter 6 on the topic of S&OP. For now, however, it is time to move down the ERP business model to the next step, operations planning. In this supply-side planning methodology, we will discuss the manufacturing

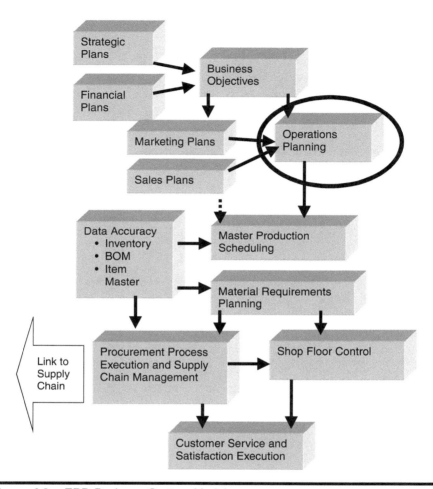

Figure 4.8. ERP Business System Model.

issues in more detail. As defined in Figure 4.8, the operations plan is a main process with many important inputs and outputs and is the other side of the necessary handshake for the prerequisite, demand planning.

5

OPERATIONS PLANNING

If you are reading this book, you likely are an operations person by trade. You may have been in operations all your life. I know this because, like you, I am a life-long operations employee and manager. I like the smell of the factory. Making things from raw materials and doing it with speed and flexibility is an art well appreciated by people like us. That is what operations planning is all about — making things with cost effectiveness, speed, and high customer satisfaction. Class A ERP process definitions in the operations planning space help to ensure predictable outcomes to schedules.

Operations planning starts with the demand plan and an understanding of the capacities and/or process expectations of the business. With this understanding of process expectations comes the necessary definition of inventory strategy. Too often, inventory strategy is not given the credit it deserves.

> Too often, inventory strategy is not given the credit it deserves.

Inventory strategy is a key design element at this point in the Class A ERP discussion. This term is used to designate how inventory is to be positioned in the overall manufacturing, supply chain, and customer service processes. At The Raymond Corporation, the overall product mix was about 80 percent make-to-order (MTO) or engineer-to-order (ETO) products and 20 percent make-to-stock (MTS) products. Having more than one strategy is pretty typical. Understanding and acknowledging it helps to put the right process in place to meet inventory goals and cost requirements.

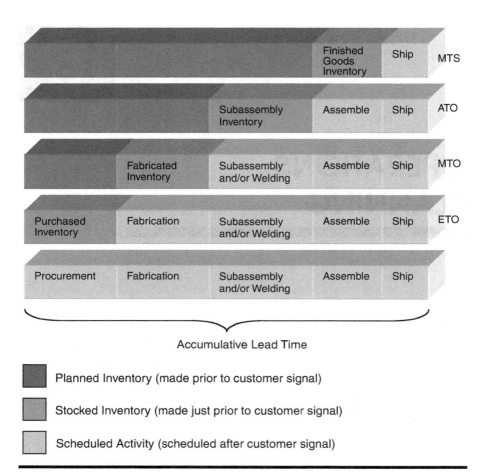

Figure 5.1. Inventory Strategy.

In MTO businesses, the inventory is usually planned at a raw material level, and the process of machining, welding, subassembling, and assembling the materials does not happen until the customer enters the order with the required specifications. A company that has received a lot of attention for its inventory strategy is Dell Computer. Its inventory strategy is to master schedule the subassemblies and complete the final assembly at the last moment when the customer signal is received. This allows the company to provide configuration flexibility specific to a customer's need and yet allows fast cycle time from customer signal to shipment. Dell's inventory strategy would be called assemble to order (ATO). Figure 5.1 gives a sense of how these strategies fit together. Keep in mind that while it may not be the case with Dell, most companies do

It is not uncommon for companies to have several inventory strategies with separate pricing and service requirements for each.

not use just one of the strategies. It is not uncommon for companies to have several inventory strategies with separate pricing and service requirements for each.

The inventory strategy is important because the planning world gets its signals and revolves around these strategies. Although many companies do not acknowledge this, *all* companies have to make these inventory strategy decisions. To make matters worse, many organizations will say they use one type of strategy and actually do something totally different. I was called into a company by a friend who had just taken a new position as VP and general manager of an operation that manufactures in Mexico but is headquartered near Los Angeles. My first visit was to the headquarters, where I met with the top-management staff, VP of engineering, VP of sales, VP of operations, and the president. During the conversations, I asked what their inventory strategy was. The president replied, "We are a make-to-order company and only a make-to-order company. We do not buy or start anything without a purchase order or schedule instructions from the customer." I found this interesting because I knew that at least one of the company's customers, Caterpillar, required a two-day lead time. I decided not to argue, but instead to wait and see for myself during the plant visit scheduled for the following day. When we arrived at the plant, I was able to talk to Dennis, the master scheduler. One of the first questions I had for Dennis concerned the inventory strategy. His take was "We are strictly a make-to-order company. We do not buy or start anything without a purchase order or schedule instructions from the customer." Sound familiar? In further discussion, it was revealed that, in fact, one of the company's most important customers was fair, but very demanding. This customer would give forecasts out several weeks in the form of a blanket order. The customer also gave (semi-firm) schedules out a couple of weeks, but ultimately and predictably gave firm schedule releases from that schedule two days prior to required shipment. The lead time for one main component in this particular process was up to sixteen weeks. This sixteen-week-lead-time purchased part was a specialized copper element made almost exclusively for this application. There was also a welding process that took a couple of days prior to assembly. Assembly took another day. All of this time added up to a questionable statement that this organization was strictly an MTO company. We all know that it can be difficult to have a sixteen-week-lead-time item with an additional three days in house and deliver product in two days using an MTO strategy. I asked Dennis about this. He replied, "Well, Caterpillar is a demanding customer. They often change Tuesday's schedule on Monday. We do not want to let them down, so we make extra inventory of the popular configurations and store it on the line in anticipation of these last-minute schedule changes." As I heard this, it became obvious that Dennis did not schedule exclusively MTO inventory strategy! Even he did not acknowledge this reality. By anticipating demand, Dennis was actually doing some MTS strategy. This is typical in many organizations. The lesson in this situation is to acknowledge the process as it exists. With the proper acknowledgment, management is more in control of the risks taken and nobody gets surprised by inventory that is built into the process by design. Inventory strategy and the effects on master scheduling will be detailed in Chapter 7. Inventory strategies change the method used for master scheduling suppliers and components. Inventory strategy is smart policy, but the rules agreed to internally need to be understood by all the players — what products, how much, and when. It also needs to be noted that inventory strategy is not designed to say no to the customer. Inventory strategy is acknowledged so that everyone in the business is on the same wavelength and it is acknowledged that there is some cost when you say yes.

> When lead-time requirements from the customer are shorter than the total accumulative lead time of the supply chain and manufacturing process, some of the process must be planned.

There is no argument that this is not rocket science. When lead-time requirements from the customer are shorter than the total accumulative lead time of the supply chain and manufacturing process, some of the process must be planned.

That means that some of the lead time in the process is committed to via forecasted requirements. Sometimes, as in the case of this supplier to Caterpillar, some finished goods were made in anticipation of the actual customer order. Inventory strategy should be mapped out, agreed to by both the demand- and supply-side management, and rules of engagement documented. When the market or strategy changes, it is important to update the handshake between the parties. Again, the objective is lowest cost with highest service. To acknowledge the realities of inventory strategy is to allow for the most cost-effective and highest customer service processing of orders. As will be explained in Chapter 7, inventory strategy makes a difference in how the scheduling will be done and what techniques will be used. The discussion on product families should also be reopened.

PRODUCT FAMILIES

Let's use an example of a company that makes automotive components. This sample company sells directly to an original equipment (OE) manufacturer,

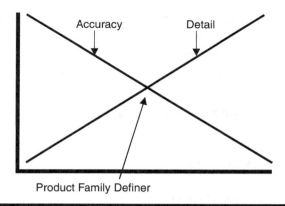

Figure 5.2. Accuracy Versus Detail Comparison (from Chapters 3 and 4).

through aftermarket distribution, and also direct to builders for custom hot-rod work. Radiators are one of the main products in this hypothetical example. We will call this company Auto Inc., and we will deal with just one product in this example even though there would most likely be several other product families. The sales forecast for Auto Inc. is in total units for the radiator end-item family. Planning capacity and component availability become complex when lead time and inventory position for each are totally different.

A better approach is to take the product family designations to one more level of detail using the inventory strategy as an important part in the decision. Auto Inc. makes many different automotive components, but radiators are one of the main products manufactured at this facility. Auto Inc. might want to have one product family for MTS radiators in anticipation of the OE orders, another product family for the MTO product for the aftermarket distribution where you expect them to carry the inventory, and a third ETO product family for the low-volume custom product supplied to the hot-rod builders. At this level of detail, in most cases, we do not violate the handshake around the lesson learned in Figure 5.2.

> Cardinal rule in Class A ERP: All top-management planning processes (business planning, demand planning, and operations planning) must be designated and measured in the exact same product family designations.

The detail/accuracy plot (Figure 5.2) is important to acknowledge. By defining families in our example to MTS radiators, MTO radiators, and ETO radiators, we also change the demand planning requirements and the business planning requirements.

That means the business planning families equal the demand planning families equal the operations planning families. This way, there is little confusion about what certain inventory scheduling methodologies cost and how much margin they are producing.

ELEMENTS OF OPERATIONS PLANNING

Operations planning has a few critical elements that should be discussed. In this chapter, we will look at each one individually.

1. Rough cut capacity planning, internal
   a. Capital equipment
   b. Factory locations
   c. Factory capacity
2. Supply chain capacity, external
3. Detailed resource planning
   a. People
   b. Skills
   c. Machinery capacity short term
   d. Plant capacity short term
4. Objectives planning in support of the business plan
5. Measurements

Rough Cut Capacity Planning

Operations planning is the culmination of all the responsibilities of the supply side of the business. There should be great respect for all successful VPs of operations throughout the world. Maybe it is because I have always been on the supply side of business, but it is my perception that they do not get the respect they deserve. Getting it right is not always a piece of cake. Operations planning is about having the right material in place when the customer needs it. Operations planning requires the right skills, the right machinery, the right capacity, and the right process at the right time — no small task (Figure 5.3).

> Operations planning requires the right skills, the right machinery, the right capacity, and the right process at the right time — no small task.

That is why the best operations professionals have sound management systems and use measurement as a window on process proficiency and predictability.

Balancing Existing Capacity and Inventory
with Forecasted and Actual Demand

Figure 5.3. Operations Planning.

Rough cut capacity planning starts with the overall strategic plan and overlaps the twelve-month rolling demand plan. Having these plans in the same shared product family designations is helpful for the planning process, especially when it comes to machinery, line, and plant capacity issues. The strategic element of rough cut capacity planning is simply the assurance that the business is set with plants and equipment for the three- to five-year plan. For example, knowing that a new plant will be brought on line in less than two years will normally affect decisions throughout the business now. Also, adding (or shutting down) plants is normally not a twelve-month process, and looking ahead is the only sane way to approach this monumental task. It is impossible to do a good job of rough cut capacity planning without linking it to the supply chain expectations.

By using both good planning process involving handshakes with the demand side of the organization along with good data gathering, the task of rough cut capacity planning becomes doable. The questions are not complicated:

- Does your organization have enough plant capacity to produce product to meet the long-term strategic plan?
- Does your organization have enough machine capacity and people to support the plant needs?
- Are the plants in the right location to meet customer demands as forecasted by the strategic plan?

There is no magic or shortcuts in this space. No one has clear vision into the future, and nothing is certain. Risk management and good business savvy are prescribed here. *Top management* must do this, however. It cannot be delegated down the organization effectively. Lastly, ignoring rough cut capacity planning is not an option in high-performance, growing businesses. Lean manufacturing requires quick reflexes. Anticipation of customer moves is very helpful in managing for flexibility.

Supply Chain Capacity

Very few companies have been able to survive, following Henry Ford's early lead, by maintaining a completely vertically designed manufacturing process flow. In Henry's early days, iron ore was taken from the mines owned by Ford and processed, machined, welded, and assembled into automobiles completely within the ownership and watchful eye of the Ford organization. This level of vertical integration is rare today. Most companies just find it too difficult to be the best at everything and rely on other specialists for some processes. Good conscience, regulations, and environmental laws have also made it difficult to finance environmentally friendly and regulation-compliant equipment amortized on just its own manufacturing requirements. It often requires multiple customers to afford the capital equipment. This has led to supply chain dependency.

To make matters worse, as most readers are aware, the introduction of a global market has complicated the equation even more. Now low-cost regions with improved quality and transportation have facilitated the movement of supply chain legs to areas very far away, sometimes (often) on the other side of the globe. High-performance organizations today are as cognizant of their suppliers' capacity as they are of their own. Critical operations and company material flow are dependent on these suppliers providing a continuous stream of parts almost regardless of what normal process variation is thrown at them. Capacity confirmation needs to happen not only through metrics but also through regularly scheduled visits and phone discussions. This needs to occur at a relatively high level within both organizations. With the top managers involved, it is more likely that there will be appreciation for critical expenditures or actions required by the people who can govern these decisions.

A certification process for suppliers is often helpful, and most high-performance organizations have such a program. These efforts focus on certifying the supplier's process at the source and auditing periodically to confirm compliance. This approach is much more cost effective than inspecting or ignoring supplier quality and predictability. This will be covered in more depth in Chapter 10.

Detailed Resource Planning

The detailed requirements for operations planning are different from the rough cut requirements in that they are shorter term and at a different level of detail. The areas of concern are (1) manpower, (2) skill sets, and (3) machinery and plant capacity. In each case, high-performance organizations will understand both expected need and existing resources as best they can. Detailed plans are developed using the sales and operations planning (S&OP) outputs, and any required additional actions are noted and followed up on.

Manpower

In the case of manpower, most organizations have a ratio (by product family) of manpower requirements per unit (Table 5.1). This is yet another reason to have all the planning done in the same product family designators. By taking the demand plan by product family and time phasing the requirements, the planner can see if there are gaps that need to be dealt with in the future. The idea is to know about these gaps in time to deal with them in a cost-effective manner. When the forecast is applied to the existing resource acknowledgment, actions can then be taken as necessary. These actions might include scheduling overtime, scheduling temporary employees for certain periods, or, in the other direction, promoting use of vacation in a certain time period or even asking for volunteers to take time off.

Skill sets are a little less mathematical, but still need to be dealt with thoughtfully for sustainable high performance. Skill is a much more important capacity requirement today than it was a few years ago. Technology has replaced backbreaking work, and computers have replaced some skills but have created requirements for new ones at the same time. For example, McDonald's does not necessarily need to have clerks skilled in arithmetic because the point-of-sale (POS) terminals are fairly foolproof. When a clerk types in a Quarter Pounder®, the cost is automatically entered and displayed. When the amount of money received from the customer is punched in, the POS terminal then tells the clerk exactly how much change to give back to the customer. Instead of arithmetic skills, McDonald's has the added need for information technology resources and skills and maybe customer service skills. Manufacturing busi-

Table 5.1.   Capacity Plan for Manpower Resources for Product Family A

	Jan	Feb	Mar	Apr	May	Jun	Jul	Aug	Sep	Oct	Nov	Dec
All Units and Hours in Thousands												
Forecast in units	43	54	55	56	78	76	65	55	51	44	54	45
Required manpower hours	215	270	275	280	390	380	325	275	255	220	270	225
Existing manpower	280	280	280	280	280	280	280	280	280	280	280	280
Delta in hours	65	10	5	0	–110	–110	–45	5	25	60	10	55

nesses have similarly migrating skills. Understanding this is important and is the responsibility of management (including first-line and middle managers in the organization). Again, the use of a forecast from the demand planning process is essential to understand the needs of the coming year. The twelve-month rolling demand plan by product family is not only extremely helpful, it is essential.

Machinery

The last discussion point in this detail resource planning focus is machinery, short term. Machinery, like people skill requirements, is changing continuously. New technology affects the flexibility, cost, and speed of operations when applied correctly. When used wisely, capital spending can help maintain competitive advantage. Keeping abreast of the latest technologies is a good idea regardless of the money available or the need for additional capacity. Sometimes, it is just smart to migrate to new technology for the sake of efficiency, even if there are no capacity issues. It is probably fair to say that business savvy and awareness are the advantage factors in this space. There is no magical answer to how much or when, but asking and answering the questions regularly will produce better results. Most companies today are using quite short payback periods for investment (Figure 5.4). Some companies still use the five-year window, but many have moved to two- or three-year payback justifications. The reasons are twofold: (1) there are a lot of opportunities in most businesses and (2) no one really has a clue as to what will happen in five or six years due to the movement of technology and market expansion.

In the detailed resource planning space, the emphasis is on shorter-term payback periods. The longer-term strategic planning focus deals with the three-to five-year horizon normally. There is a different thought process for each. Detailed planning is much more about the day-to-day running of an operation and is linked to the demand plan agreed to and blessed by management in the S&OP process review. Three- to five-year planning is integrated with market

1 year — Very ideal payback period for most business investments	2 year — Ideal payback period for most business investments	3 year — Reasonable payback period for most business investments	4 year — Long payback period for most business investments	5 year — Maximum payback period for most business investments

Figure 5.4. Capital Payback Period for Most High-Performance Businesses.

and process changes planned by management in the strategic space. While they are always linked, there is a need for a separate focus in both arenas.

Along with the detailed planning discussion on machinery is the plant capacity, short term. Linked with all of the other points, it is still important to review the overall short-term outlook for capacity. When looking at the long-term plant capacity in the strategic arena, there needs to be a 40,000-foot view down to the plants, including location, size, and quantity. In the detailed view of the horizon, again there should be twelve rolling months. The view in this space is more internal to the organization. This is the planning space where things like shifts, number of people, machinery capacity short term, and use of temporary employees are evaluated and decisions made (Figure 5.5).

Objectives Planning in Support of the Business Plan

In the discussion surrounding the business plan in Chapter 3, the elements of goal setting were described. The strategic goals and business imperatives are delegated to the various departments within the organization. The operations division usually ends up with several of these important goals. In high-performance organizations, these corporate objectives are the lifeblood of performance measurement and/or bonuses. Remember that the definition of business imperative is that "*it will happen in the next twelve months.*" These imperatives must be of the highest-prioritized activity.

High-performance operations groups have predictable or repeatable reviews of corporate or company strategy to ensure proper linkage to the objectives within the division. There is often a similar approach at the division level to the company or corporate level. Normally, there are both operations division strategic objectives and business imperatives. The objectives either directly link or are the same as the corporate objectives and imperatives. Figure 5.6, which depicts this relationship, is repeated from Chapter 3.

The model in Figure 5.6 is still appropriate within the operations planning hierarchy. The only difference is that objectives are limited to those specific to the operations division. The operations equivalent hierarchy might look more like Figure 5.7.

The hierarchy is essential to keep the objectives linked properly. Choosing the right objectives may be the real highest value-add process initially. Choosing is not always a mistake-proof exercise, but without a Class A or Six Sigma type management system, even if your savvy has allowed you to pick the right objectives, value is not always harvested efficiently or at all. Class A ERP and/or Six Sigma project management skills and management system structure work well to integrate goals with good project management process. Six Sigma will be described in detail as it integrates with Class A in Chapter 13.

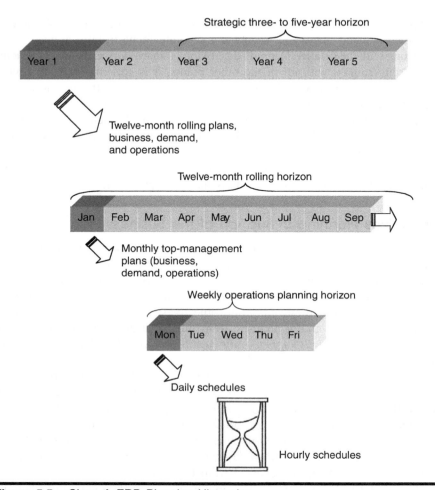

Figure 5.5. Class A ERP Planning Hierarchy.

Operations should meet at least once a year to plan these objectives and align with the corporate vision. In many companies, this starts with an analysis just prior to the corporate strategic session. This helps to ensure that operations management is prepared for the top-level assessment coming. A postsession analysis is appropriate after the corporate strategic session has been completed and decisions made at that level. It is then time to get operations back together to evaluate current position in relation to any changes that may have happened at the corporate level. It also gives the VP of operations a good chance to update the rest of his or her staff on any pertinent discussions from the corporate

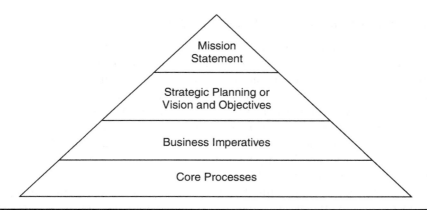

Figure 5.6.   Business Planning Hierarchy.

meeting. If all of these strategy and objectives meetings are held on the same day each year (such as the third Thursday in August, for example), no one misses them because of vacation, is caught off guard, or is not prepared for the meeting when they get there. The second meeting (after the corporate strategic session) is the gathering where the operational business imperatives are determined. Sometimes these are simply carryovers from the corporate session, but sometimes there are additional imperatives that need to be addressed to meet other needs specific to the operation. These might include training, additional capacity investments, or transferring operations from one plant to another for efficiency or cost reduction.

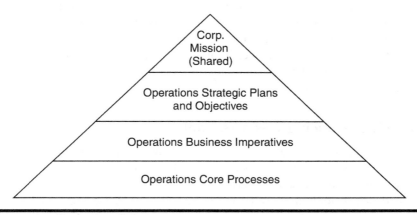

Figure 5.7.   Operations Planning.

Role of Operations in the S&OP Process

The role of operations in the S&OP process is significant. In most organizations, because of the master scheduler's role, the involvement of operations would be considered a leadership role in the success of the S&OP. Chapter 6 will get into much more detail on the support required from the operations group to establish high performance. With that in mind, it still needs to be said that the main responsibility and expectation from the function are capacity, quality, low cost, flexibility, and responsiveness. This can obviously be a challenge! Class A ERP integrated with lean and Six Sigma delivers this. Master scheduling, shop floor control, and project management will provide additional pieces of the puzzle. The S&OP process is the culmination of the full forces of operations represented in a commitment that covers costs, inventory, and customer service. Realistically, operations has a great load to carry, and when Class A ERP processes are in place, it is delivered predictably.

Process Ownership in Operations Planning

Process ownership in operations is not difficult to determine. The top manager in the division is the process owner. This is usually the VP of operations. In larger companies where there are separate operations for Asia, South America, Europe, etc., there often is a separate Class A initiative in each region. This leaves process ownership to the ranking operations manager in each region. Process ownership means taking responsibility for the metric and process performance. It would include accuracy of capacity plans committed to and executed monthly. These plans of record for measurement purposes would be documented in the monthly S&OP review.

In Class A, there are many more processes subservient to the operations planning process, and each has a specific ownership and accountability. These are not to be confused with the top-management planning process ownership. Accountability of the top manager in operations happens at the CEO level and is evident at the S&OP review and other project management and performance reporting events at the top level.

Operations Planning Measurement

The measurement in operations is simple and straightforward. As you can see in Table 5.2, the metric is strictly evaluated by product family. The product families must be the same ones that are used in the demand and business plans. The tendency to look at overall units produced is good for trend analysis and tempting for measurement purposes, but is not in the spirit of the Class A ERP

Table 5.2. Operations Planning Metric

	All Units in Thousands		
	S&OP Commitment in Units	Actual in Units	Performance
Product family A	90	92	98 percent
Product family B	50	49	99 percent
Product family C	78	78	100 percent
Product family D	8	8.5	94 percent
Product family E	155	157	99 percent
Total performance			98 percent

effort. Process linkage is, after all, the reason why product families have to be common throughout the planning process.

The purpose of the metric calculation is to find the average accuracy per product family. Some mathematicians and engineers will get nervous about this approach as it is realistically an "average of averages." It is, but this is the way it needs to be calculated. The calculation in this example is: (98 percent + 99 percent + 100 percent + 94 percent + 99 percent)/5 = 98 percent. Notice also that the metric recognizes an absolute deviation from plan. In product family D, 8,500 units were produced against a plan of 8,000 units. A similar situation occurs in family E. There was a reason for this miss, and the metric points toward finding that reason. To get to Class A ERP performance, both a good plan and good execution are necessary.

> To get to Class A ERP performance, both a good plan and good execution are necessary.

When the plan is outperformed, sometimes it is simply bad planning. When the metric points out inaccuracy, it can be addressed and the organization can learn from it. This learning is valuable. The spirit of Class A ERP is to have predictable processes and performance. Cost and inventory levels are usually critical for meeting company objectives. It is important that both plans and execution are accurate.

I may regret sharing this next paragraph, because I am about to reveal a trade secret, at least from my viewpoint. You may be saying to yourself as you read the metric calculation that in *your* business, product families are so unequal in demand and volume that your recommendation will be to *weight* the values in line with monthly volumes. In effect, the result will no longer be the "average" accuracy per product family. Here is my trade secret, and if you are the Class A ERP champion or team leader, I would recommend you heed what my experience has taught over the years. In the beginning of the Class A implementation, **do not** weight the values. It only hides lessons that benefit the

Remember that the objective is high performance, not "95s in boxes."

organization long term. Remember that the objective is high performance, not "95s in boxes."

If that is your attitude, you will see the point of this argument. If the organization is serious about root cause and uses the metrics to learn and drive change, the lessons that are learned will be valuable. After the performance is in the mid-90s and the business is in a much more sound position from the standpoint of understanding process variation and reasons for inaccurate plans, there is less risk in playing around with the numbers and weighting is more appropriate. During Class A ERP implementations, I frequently will insist on straight averages until the team has worked the root causes to a reasonable level of understanding and performance. It is a valuable learning experience worth going through.

There *can* be exceptions to this rule. If there are, for example, small product families that truly are differentiated from the rest of the normal build, there may be a need for weighting from the start. For example, at one client company, there is a kitting operation that shares no capacity or components with other operations. It is truly a stand-alone, but small, process. This product family represents less than 1 percent of the business. Five other product families share the balance of 99 percent, ranging between 8 and 33 percent. The only family that is weighted on volume is the small kitting family. The rest would have 99 percent divided evenly on weight in this example. Whatever you do in this regard, make sure you go after the root cause of each family and eliminate the reasons for process variation. The kinds of process interruptions experienced in one family, even a small one, will usually be helpful in the whole business. The more weighting is applied, the less effort will be applied to the small families. Do not fall into this trap. Enough said.

Operations planning is arguably the heartbeat of any manufacturing organization. Although the sales department will argue that its importance is second to none, it is pretty difficult to meet customer need without operations. Getting this process right is worth the effort.

6

SALES AND OPERATIONS PLANNING

The sales and operations planning (S&OP) process is one of the most talked about and appreciated process topics in business today. The interesting aspect about this is that it is not a new process. In fact, some might agree that it is an old process. There is just a lot of new interest in it, and for good reason — it pays off. Tim Frank, CEO of the Grafco PET packaging company, a well-managed blow-molding business headquartered in Baltimore, has become so convinced of this that his company is working with its customers to coach improvements in their S&OP processes. The company actually has high-level managers who regularly attend customer S&OP processes just to help with improving the mechanics. Many organizations are working with their suppliers in this same mode, which is more common.

Virtually all high-performance organizations do some form of an S&OP process regularly. In these organizations, the S&OP is a monthly top-management planning review meeting where metrics and performance are assessed and adjustments are made based on recommendations from data collection and analysis done in preparation for this review. The key to success is preparation and good data mining in advance of the decision process. This activity is followed by an analysis of the "30-60-90"-day time frames going forward. Beyond ninety days, only the exceptions are reviewed through the rest of the twelve-month rolling horizon. A complete list of agenda components within the S&OP follows.

1. Review of last thirty days — The accuracy metrics by product family are reviewed.
   a. Financial plan accuracy by product family — The top-management financial manager (usually the CFO) reports the performance number for the financial plan accuracy and communicates root cause analysis of any process variation and actions to improve upcoming accuracy.
   b. Forecast plan accuracy by product family — The top-management demand-side manager (usually either the VP of sales or the VP of marketing) reports the performance number for the demand plan accuracy and communicates root cause analysis of any process variation and actions to improve upcoming accuracy.
   c. Production plan accuracy by product family — The top operations management manager (usually the VP of operations) reports the performance number for the production plan accuracy and communicates root cause analysis of any process variation and actions to improve upcoming accuracy.
2. Review of thirty- to sixty-day plan expectations — Risks and/or changes since the last monthly plan are reviewed in detail.
   a. Financial plan risks reviewed by the team — This is normally led by the CFO, but in this section, the CEO and other staffers will have inputs.
   b. Sales forecast risks reviewed by the team — The demand plan process owner is the facilitator in this section of the S&OP. The CEO is often the tie breaker in this space if there are varying confidence levels. Disagreement can happen if recent forecasts have been inaccurate.
   c. Production risks — The operations process owner presents any risks in the near term and actions planned. This can also be a lively discussion.
3. The 90- to 120-day plan expectations — This time frame is a cursory review, and depending on high-risk situations in the future, this can be either a short conversation or one that gets into a lot of details. Some of the more common risks are:
   a. New product introduction is discussed for risks of plan accuracy.
   b. Production shifts to alternative sources such as moving product lines from one internal plant to another, migration to offshore sights, supply chain risks, etc.
   c. Promotions, shows, customer actions, etc. are reviewed on an exception-only basis.
   d. Review/verification of normal cyclicality/seasonality is done.

  e. Anticipated currency exchange issues affecting the plan accuracy are discussed.
4. The balance of the twelve-month rolling schedule is quickly reviewed for anomalies or changes expected.
  a. Any new information is communicated.
  b. Changes from the previous plan are reviewed.
  c. Any effects to the year-end numbers are normally reviewed.

PARTICIPANTS TO THE S&OP AND THEIR ROLES IN LARGER MULTIPLANT ENVIRONMENT

There can be a few variations in participation and roles depending on the organizational structure and size of a business. Two views will be used as examples in this template, the first of which is a larger multiplant environment.

In a larger organization, often there will be product managers responsible for overall product success, including creation of demand, product cost, and overall profitability. It is a highly responsible job, and even though they are not always considered top management, the managers really are in the solid core of people who influence the planning process. Larger organizations also usually have a demand manager within either the sales or marketing leg of the business. In some organizations, the demand manager reports to the VP of sales, in some the VP of marketing, and in yet others the VP of sales *and* marketing; again, it depends on the size and scope of an organization. In this example, we will assume that both positions are held by one VP, as well as the normal expectation in most businesses of a CEO, CFO, VP of operations, and a master scheduler. Figure 6.1 depicts the S&OP participants under these assumptions. Note that in this example, there is no demand manager; the forecasting process would be executed through the product managers.

In smaller organizations, there are generally fewer top managers, each of whom wears multiple hats. Because of this, participation in the S&OP will engage a smaller group, but the duties are all still executed. In smaller businesses, there may not be a product manager for each product family, but there might be a demand manager. The demand manager normally reports to the ranking demand-side top manager, such as the VP of sales and marketing The demand manager collects data from the field and consolidates input information to develop and maintain the demand forecast. Figure 6.2 shows the typical S&OP participation in such an organization.

Very small organizations do not have demand managers. This does not stop the process, but the duties have to be picked up by the sales or marketing

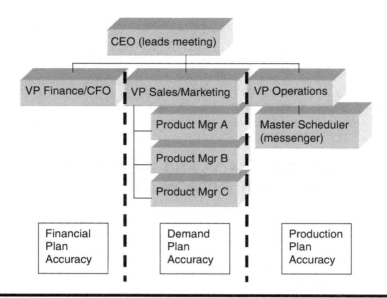

Figure 6.1. Typical Participation in the S&OP: Larger Businesses.

administration. In some organizations, the responsibilities are filled by having an administrative assistant pick up the clerical pieces and the VP of sales and marketing do the analysis. None of these scenarios is better than another; it

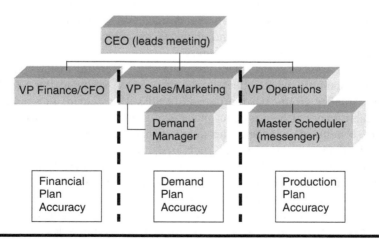

Figure 6.2. Typical Participation in the S&OP: Smaller Businesses.

simply depends on the complexity of the market, number of products, appetite of the VP of sales, and size of the organization.

SOME THOUGHTS ON ALTERNATIVE LABELS FOR S&OP

In some organizations I have worked with, the term S&OP has an additional letter. This is prevalent enough to include the acronym "SIOP" in this text. SIOP (sales, inventory, and operations planning) is synonymous with the S&OP process. Similarly, PSI (production, sales, and inventory) is another acronym used in this space. In my experience, organizations that have picked up the PSI version and abbreviation tend to share a lower evolution of S&OP DNA and are often missing some of the important elements of planning and accountability. There is nothing scientific in this observation, and very little has been written about the ancestry of the various terms beyond S&OP, the most popular title for the popular top-management planning process. Most of the time, when companies refer to these alternative process labels, they are speaking of the top-management planning process.

ROLES IN THE S&OP MEETING

The monthly S&OP meeting in a high-performance organization will have a very predictable agenda and clear roles for the players. First of all, the meeting belongs to the CEO or president, whoever is in charge regularly day to day. This monthly meeting is a main element of the Class A ERP management system, so accountability needs to be a top priority. Accountability does not mean that bloodletting is required. Instead, it means that process owners come into the meeting prepared, with their homework done, and present the facts, including actions to offset process variation. Accountability is best ensured if there is someone to play the role of asking the tough questions. That role resides with the chair of this meeting, the president or CEO. Each of the other plans has a process owner, and it is the role of each of these owners to propose updates or changes to their plans to help eliminate plan inaccuracies. The duties of each person in the meeting are described in the following section. Some alternative participants are also listed. Each organization will be a little different in its choice of structure. There

> There is no right or wrong organizational chart as long as the right process owners are accountable for the accuracy and facts relating to their plans.

is no right or wrong organizational chart as long as the right process owners are accountable for the accuracy and facts relating to their plans.

Duties of Each Process Owner in the Monthly S&OP Meeting

1. President/CEO
   a. Maintains the meeting schedule twelve months in advance (usually done by always holding it on the first Tuesday of every month, the second workday of every month, or some other predictable schedule).
   b. Leads the meeting.
   c. Makes sure the meeting is not preempted by some other priority.
   d. Ensures the priority of attendance so process owners all show up regularly.
   e. Monitors and facilitates a consistent agenda.
   f. Asks the tough questions as metric performance and actions are presented for each plan (financial plan accuracy, demand forecast accuracy, operations plan accuracy).
   g. Minutes are normally published from his or her office.
2. VP of finance/CFO
   a. Prepares the financial spreadsheets for the S&OP.
   b. Presents metric performance numbers for financial plan accuracy, by product family, for the past thirty days.
   c. Shows trends in product family financial plan accuracy, month to month, for at least the last six months.
   d. Shows delta between plan of record and the last product family financial forecasts.
   e. Presents actions to close any accuracy gaps.
   f. Updates and shares product family financial plans going forward, "30-60-90" days and balance of twelve-month rolling plan.
   g. Highlights any changes affecting promises to stakeholders.
3. VP of sales and/or marketing
   a. Presents metric performance numbers for demand plan accuracy, by product family, for the past thirty days.
   b. Shows trends in product family demand plan accuracy, month to month, for at least the last six months.
   c. Shows delta between demand plan of record and the last product family forecasts.
   d. Presents actions to close any accuracy gaps.
   e. Updates and shares product family demand plans going forward, "30-60-90" days and balance of twelve-month rolling plan.

    f. Highlights any changes affecting forecasts, such as promotions, large shows, phase-outs, new product introductions, etc.

    g. Answers the questions in the meeting concerning the forecast accuracy.

4. VP of operations

    a. Presents metric performance numbers for the operations plan accuracy, by product family, for the past thirty days.

    b. Shows trends in product family operations plan accuracy, month to month, for at least the last six months.

    c. Shows delta between plan of record and the last product family operations forecasts.

    d. Presents actions to close any accuracy gaps.

    e. Updates and shares product family operations plans going forward, "30-60-90" days and balance of twelve-month rolling plan.

    f. Highlights any changes affecting plans, such as supplier risks, production shifts from one facility to another, and/or any other unusual circumstances.

    g. Answers the questions in the meeting concerning the operations plan accuracy.

5. Master scheduler

    a. Prepares the demand and operations plan spreadsheets and performance metric performance, by product family.

    b. Distributes the demand and operations plan spreadsheet to the S&OP participants prior to the meeting (as far in advance as possible).

    c. Meets with the CEO/president prior to the S&OP meeting to make sure he or she knows all of the issues and can ask all of the right questions concerning upcoming issues.

    d. Is not at the meeting to answer operations questions, although the master scheduler can be of assistance to both the operations and demand plan process owners in the meeting by providing data and facts.

6. Other possible S&OP meeting attendees:

    a. President's administrative assistant

        i. Takes notes during the meeting, including any assignments or agreements for action.

        ii. Publishes the minutes after review by the CEO.

    b. Demand manager (if the organization has one)

        i. Prepares the demand spreadsheet instead of the master scheduler.

        ii. Ensures that the demand and operations spreadsheets are in the same format and are in complete synchronization.

        iii. Provides facts and data in regard to the demand plan.

    c. Product managers (if the organization has them)
       i. Provide forecasts for their product families.
       ii. Present actions affecting customer behavior.
       iii. Answer the tough questions regarding plan accuracy of their product families.
       iv. In many organizations, these people share full accountability with the VP of sales/marketing for plan accuracy. They normally report to the demand side of the organization.
    d. VP of engineering — The VP of engineering is often a valuable attendee at the S&OP process meeting. This process owner is usually well served to understand the priority and issues around risks in upcoming schedules as well as longer term. New product introduction typically is one of the biggest risk management opportunities in the planning horizon. Engineering obviously has a big role in its successful plan accuracy.

TIMING FOR THE S&OP PROCESS AND THE "MONTHLY CLOSING OF THE BOOKS"

The S&OP process needs to happen as quickly as possible each month. In many high-performance organizations, this means that the meeting is always held on the second or third workday of the month. This makes last month's information very fresh and allows decisions to result in changes that can affect the current month while there is still time to make a significant difference. Some organizations have decided to hold it later in the month. The normal reason given is waiting for the books to close. The best organizations do not worry about the books being closed. In fact, there is an advantage for them not to be. After all, this meeting is about plans and planning accuracy. When the demand and operations plans are accurate, the financial plans are often equally accurate. When these plans are not accurate, there is a great learning opportunity. If you are from the financial department and are wondering how the results can possibly be reviewed at the S&OP if it is done prior to the books being closed, you need to recalibrate your paradigm for this model.

The financial results should be tied directly to the demand and operations plans. If, for example, the operations spreadsheet planned certain costs by product family, producing more or less units would affect cost totals. This can be estimated quite closely if the fixed and variable costs are separated. (Fixed costs are things like heat, property taxes, some portion of utilities, some indirect labor support activities, etc.) This modeling can get as sophisticated or complex as you want it, but the simpler models seem to work the best as they are more

Table 6.1. Applying the Financial Impact to the S&OP Spreadsheet

Assumptions:

a. Product line A is make-to-order (MTO) product family.
b. In product family A, the cost per unit/cost of sales including labor, burden, material, and other overhead is approximately $1,000.
c. In product family A, the per-unit average revenue based on planned mix is approximately $1,895.

	January Example			
	Actual Units	**Planned Units**	**Actual Revenue**	**Planned Revenue**
MTO demand	100	115		
MTO operations	115	115		$217,925
Gross margin	$102,925	$102,925		

Gross margin = (115 × $1,895) − (115 × $1,000) or $217,925 − $115,000 or 102,925

easily understood in the S&OP discussions (Table 6.1). When plans are not made because of volume too low to offset burden costs, it can be easily explained. This is often more palatable than complex models that increase fixed costs automatically below certain volumes and have to be explained every time they come up in discussion.

It is helpful and quite appropriate to calculate this model for each of the twelve rolling months. Of course, the "actual" fields would only be populated for months past. By building a spreadsheet that includes costs and revenue estimates per product family, the job of planning becomes much easier. This again raises the discussion about product families. It is helpful to apply the following rule when appropriate: Price and/or margin differences can create the need for additional product families when bands in price or cost are too broad. Like all topic ranges, if the bandwidth in categorizing the families is too wide, the labels can become less meaningful/valuable. The general rule is to keep the number of product families under ten, with six as an ideal number. Having said that, there are many examples of businesses that have built successful S&OP processes with more product families than the preferred six. Just use some common sense to keep it as small as logically possible.

The timing for the monthly cycle of Class A–related activities in S&OP preparation includes the demand review, the pre-S&OP, and the S&OP meeting itself (Figure 6.3). The demand review normally happens every Friday except for the last Friday of the month. On the last Friday, the demand review is replaced by the more detailed and encompassing pre-S&OP meeting. The pre-S&OP covers all the demand review topics and includes a more in-depth review

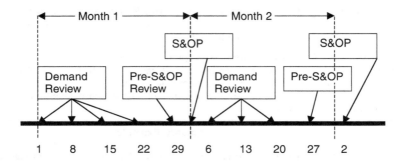

Example of possible Friday calendar dates in two monthly cycles

Figure 6.3.   Timing of the S&OP Cycle.

of both performance and risk analysis going forward. Later in this chapter, the agendas of both will be covered.

The timing of the S&OP process is important. It needs to be timely and repeatable/predictable. If it is done at the same time each month, no one will miss it with an excuse that they did not get the communication. If it is done very close to the beginning of the month, this perfect timing allows decisions to affect the current month.

AGENDA FOR THE PRE-S&OP

The pre-S&OP is a meeting held just prior to the S&OP meeting. Since it would not be acceptable to show up to the S&OP without root cause analysis completed and proposals for continued improvement prepared, it makes sense to have a process owner gathering just prior to the main top-management meeting to determine issues and analysis actions as required. A company CEO deserves to see recommendations and action plans at the S&OP, and that is exactly what gets accomplished at the pre-S&OP.

The players with required attendance are:

1. Product managers (if the organization has them); product manager alternative: demand manager
2. Demand manager (if the organization has one); demand manager alternative: product manager or VP of sales and/or marketing
3. Sales manager(s)
4. Master scheduler
5. Plant managers often attend via conference call

The discretionary attendees are:

1. VP of operations
2. VP of finance/CFO
3. VP of sales and marketing (if they are well represented by a demand manager or product manager)

The master scheduler is normally the facilitator at this meeting. The agenda is always the same, allowing for players to come prepared with actions and proposals to present. Decisions will be made and handshakes will happen, but each player should anticipate these needs and come prepared for them. Some of the kinds of decisions that have to be made include inventory plan changes, forecast changes, capacity increases or decreases, changes to promotions due to availability of product, etc. These decisions can become pretty involved, and it is helpful to the business to get the data and facts out in the open for full disclosure. The better the job that is done at the pre-S&OP, the faster and more efficient the S&OP process will be. Think of the pre-S&OP as the analysis gathering and proposal development meeting, and think of the S&OP follow-up as the decision-making meeting where the proposals are either authorized or adjusted by the top-management team. High-performance organizations get very good at preparing for the S&OP by anticipating questions that might be asked in advance and having the answers either ready or presented in anticipation of the questions.

> The master scheduler is normally the facilitator at this meeting.

The agenda for the pre-S&OP is basically the same agenda as for the S&OP, just at a different level of preparedness. In the pre-S&OP, there are unanswered questions. In some instances, this meeting is the first time these players have all been in a room together to discuss these topics since the last meeting a month ago. At the end of the day, however, the topics are the same as the final top-management planning meeting that follows this one by a few days. Those review steps in sequence are: review the financials, review the sales forecast, and review the operations plan. The sequence of detail would be:

1. Review last thirty days of performance
   a. Demand plan
   b. Operations plan
2. Review changes required going forward in current thirty days
   a. Demand plan
   b. Operations plan

**Table 6.2. Typical S&OP Spreadsheet — Product Family A
(Four Months of a Thirteen-Month Spreadsheet)**

	Act.	Plan	Perf.	Act.	Plan	Act.	Plan	Act.	Plan
Demand plan	115	100	85 percent		105		109		115
Operations plan	105	100	95 percent		100		110		115
Financial plan	$109	$105	100 percent		$105		$109		$115

3. Review the "60-90-120"-day forecast for risk management and new issues
   a. Demand plan
   b. Operations plan
4. Briefly discuss any issues longer term. This is a much shorter topic here than in the S&OP. The main focus in this meeting is the 30- through 120-day horizon.

THE SPREADSHEETS FOR BOTH THE PRE-S&OP AND THE S&OP PROCESSES

The format for the typical S&OP is pretty consistent business to business. There is a separate spreadsheet for each product family, and the history has to go at least thirty days prior to the current review period (Table 6.2).

Additionally, some organizations use a step chart or what is sometimes called a waterfall spreadsheet to see the progression of plans over time (Table 6.3). This can be done by keeping the twelve-month rolling plan intact and adding the new plan to the bottom of the spreadsheet each month. Note that for the sake of space, this spreadsheet does not show the full twelve-month rolling horizon required by Class A ERP criteria. If this waterfall spreadsheet is used

**Table 6.3. Step Chart in the S&OP Format
(Full Spreadsheet Would Run Twelve Months)**

	Jan	Feb	Mar	Apr	May	Jun	Jul	Aug	Sep	Oct
Feb plan	perf	plan	plan	plan	plan	plan	plan	plan	plan	plan
Mar plan		perf	plan	plan	plan	plan	plan	plan	plan	plan
Apr plan			perf	plan	plan	plan	plan	plan	plan	plan
Mar plan				perf	plan	plan	plan	plan	plan	plan
Apr plan					perf	plan	plan	plan	plan	plan

regularly, top management can get additional insight into changes in plans through the year. In this application, for example, one could look at August and see how the August plan has changed month to month. Sometimes this is helpful to understand the dynamics of the planning process.

SUMMARY: KEYS TO THE SUCCESS
OF THE S&OP PROCESS

The S&OP process is very important to the success of any manufacturing business. If no other section of this book is implemented at your facility, this is the one that must be. There are a few keys to remember for high-performance payback:

1. Meetings must be scheduled twelve months in advance. Use easily understood time frames, such as every Friday for demand reviews, the last Friday of the month for the pre-S&OP, and the second workday of the month for the S&OP.
2. The meeting should not be preempted for any reason short of an unscheduled catastrophic surprise or a holiday. In the case of a holiday, the meeting should be the next workday.
3. No excuses for nonattendance are acceptable. Of course, all rules are made to be broken, and short of sickness, serious personal issues, vacation, or an exception-only customer conflict, all top managers are to attend the S&OP. Some companies have allowed salespeople to miss any time they want. This behavior does not favor high performance. When key players are missing in these sessions, the meetings will not promote proper accountability or achieve the true handshake between the demand side and supply side of the organization. The demand review has a little different roster, but the same rules apply. When on those rare exceptions when there is an absence, it is expected that a fully authorized backup will attend in the missing player's place.
4. The spreadsheet needs to include all the plans: financial, demand, and operations. The master scheduler normally is the facilitator of the format, plans of record, and metrics, especially for the demand and operations plans.
5. Keep the agenda consistent meeting to meeting.
6. The first thing to review in the meeting is the performance number (percentage of accuracy). When the performance number is reviewed first, there is a sense for good or bad that is not always apparent when

just talking about the actual volume numbers, actual versus plan. For example, sometimes 460 out of 500 may sound okay, but when there is an acknowledgment that 95 percent is the threshold of acceptability, 92 percent is no longer acceptable performance and is quickly recognizable as such. The organization needs to be tuned into 95 percent minimums for schedule and data accuracy performance. Anything less is just unacceptable.

7. The VP process owners need to answer the tough questions in the S&OP process review meeting. Organizations that completely delegate the accountability to lower levels and then sit back and blister these second-tier people in the S&OP meeting do not reach the same levels of proficiency in their businesses. When VP-level process owners take the time and interest on a frequent basis to understand the risks and are willing to coach the management of them, there is always a business benefit. Some would argue that is their job anyway.

> Organizations that completely delegate the accountability to lower levels and then sit back and blister these second-tier people in the S&OP meeting do not reach the same levels of proficiency in their businesses.

8. The president/CEO has to have a very clear interest in not only the financial numbers and performance, but also the underpinnings of unit performance. The top manager must be willing to get into enough detail to understand risk management of the plans. Things like production transfers from one facility to another and new product introductions are of special concern.

> The top manager must be willing to get into enough detail to understand risk management of the plans. Things like production transfers from one facility to another and new product introductions are of special concern.

9. The master scheduler should have access to the top manager prior to the S&OP meeting. The VP of operations must also understand that this exposure and preparation for the top manager (CEO/president) to the master scheduler's information will make the process that much more valuable to the business. To appease protocol, sometimes the VP of operations goes to this prep meeting with the master scheduler and top manager. Because

> Because of this required relationship with the top manager and because VP process owners should be answering their own process-related questions during the review, it works best if the master scheduler is not frequently asked for answers in the S&OP meeting.

of this required relationship with the top manager and because VP process owners should be answering their own process-related questions during the review, it works best if the master scheduler is not frequently asked for answers in the S&OP meeting.

10. Actions are always a result of discussions during the S&OP meeting. The CEO's administrative assistant is an excellent person to take notes during the S&OP meeting and to publish the minutes from the meeting.

11. Review last meeting actions first before starting the normal S&OP agenda items.

12. Performance metrics from the top-management plan accuracy should be posted in the factory and office areas in a visible place. The names of the VP process owners should be listed with the performance percentages.

13. Celebrate successes when appropriate. This is especially important and is often a neglected component of good management process. Have some fun!

Businesses that follow these rules and put honest effort into reaching genuine handshake agreements between the demand- and supply-side operations enjoy great benefit. The S&OP process is a key element of the Class A ERP certification and, even more importantly, the payback from it. In Chapter 7, the focus will return to the ERP business model. Master scheduling is the first operations planning process below the S&OP processes and is the heartbeat of ERP application.

7

MASTER SCHEDULING

There is no more important topic within the Class A ERP than master scheduling. While the sales and operations planning (S&OP) process is definitely in competition within the ERP process for influence and impact, the master production schedule (MPS) wins from my perspective simply because without good master scheduling, the S&OP process would be no more valuable than a rain dance with little real influence over the weather. In a Class A plant, the two most influential positions are the plant manager and the master scheduler. The master schedule defines the activities required to meet the top-management S&OP, but also dictates the requirements from customer demand. This already sounds like quite a job and it is. Figure 7.1 depicts the relationship of the master schedule with the rest of the ERP process flow. Top-management planning decisions feed into the MPS and drive emphasis and create the forecasted demand. The MPS, in turn, feeds back vital data concerning customer activity and coordinates the interaction of the two inputs. This feedback affects the next S&OP cycle.

THE MASTER PRODUCTION SCHEDULE

The MPS is the detailed schedule that resides in the ERP business system and drives the inventory strategy, supply chain actability, inventory levels, customer service, and machine and capacity utilization. The main data categories the MPS drives fall into two areas: *known* and *unknown* requirements. This may seem like a simple thought, but understanding this concept and grasping the importance of its impact is the essence of understanding the master scheduling process. In Figure 7.2, the shaded area represents the known requirements.

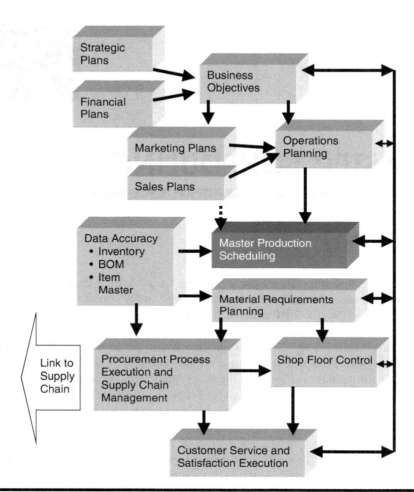

Figure 7.1. ERP Business System Model.

Where you see the shaded areas, there are many specific detail requirements driven from either firm orders from customers or firm orders from planned requirements. In the case illustrated in Figure 7.2, if every column represents one day, the conclusion from this figure would be that backlogged orders only go out about six days and even then not the full six days. Only Monday and Tuesday are close to being filled (Tuesday might even be a little over capacity per the diagram). If this company has customers that need product in less than a week and it manufactures an assembly with a cumulative lead time of more than six weeks, we would need to understand how the shipment could happen within the customer's required

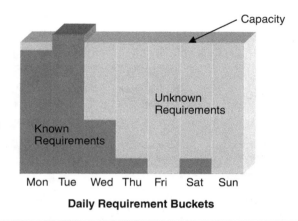

Daily Requirement Buckets

Figure 7.2. Master Schedule Known and Unknown Requirements.

time without a full six weeks, given backlog and lead time. The answer is simple and comes back to inventory strategy as discussed in detail in Chapter 5. Inventory strategy, also known as order fulfillment strategy, is more about where you plan inventory than about filling orders. When giving a class at NCR in Atlanta, I used the term "inventory strategy," and one of the students said, "Oh, now I get it. It's where in the process you plan for process buffer." Sometimes it is the little things that are the most meaningful. I guess that student was right. It is a simpler concept when you realize it is all about where the inventory is buffered in the manufacturing process. The MPS cannot be executed properly without that piece of information. In fact, the MPS and inventory strategy are attached at the hip, so-to-speak. Figure 7.3 was also used in Chapter 5 to explain inventory strategy. It works as a great illustration to show where the MPS intersects with the inventory strategy.

To illustrate the master scheduling level within the bill of material (BOM) on top of the diagram, the resulting illustration would look like Figure 7.4. At the lightening indicator in the illustration, you will find the BOM level or stockkeeping unit (SKU) at which the MPS focus would drive planned requirements.

If the organization utilizes engineer to order (ETO), the MPS *could* master schedule the engineering time and resource. In one company I worked with that manufactured capital machinery for use by the big automotive manufacturers, the master scheduler planned capacity within the engineering group. Engineering was the limiting factor and the most critical constraint in this ETO business. By master scheduling the engineering resource, the right number of engineers would be retained and strategy would be developed for subcontracting design work as required. Engineering, although a professional need, is, after all, still just one of many skilled requirements in the manufacturing process flow.

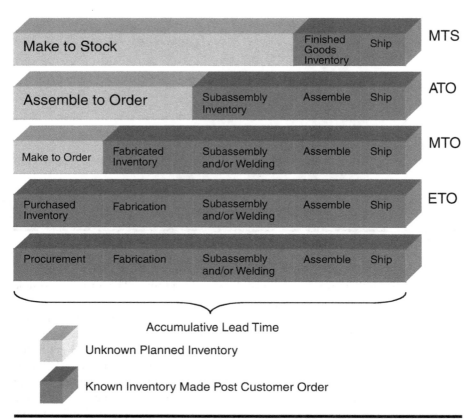

Figure 7.3.   Inventory Strategy.

The SKU level to be scheduled by master scheduling is determined by the BOM. Say, for example, that the BOM structure looked like Figure 7.5 for the inventory strategy depicted in Figure 7.4. If the business determined its inventory strategy for Assembly 123's product family to be make to stock (MTS), the master schedule for that product family would drive requirements for the top-level parts in that family, in this case Assembly 123. This means that if you went into the MPS to look at the specific part numbers scheduled, you would see the part number Assembly 123 in the scheduled requirements going forward even though there was no specific customer order for the requirement. The plan would be driven from forecasted requirements. That is the characteristic of an MTS environment.

If the strategy was assemble to order (ATO), the subassemblies and/or weldments would be driven by the master schedule and available for the cus-

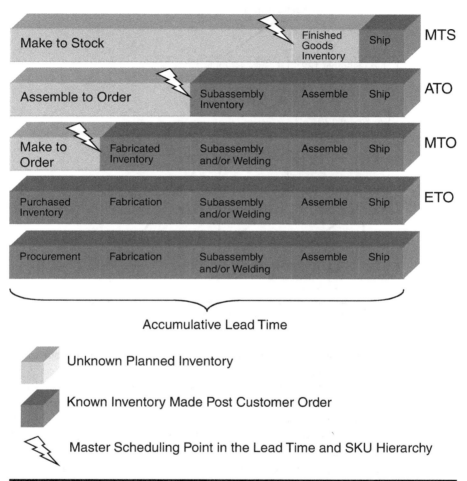

Accumulative Lead Time

Unknown Planned Inventory

Known Inventory Made Post Customer Order

Master Scheduling Point in the Lead Time and SKU Hierarchy

Figure 7.4. Master Scheduling within Inventory Strategy.

tomer order when it was received. No final assembly would happen until the order was received from the customer. The assembly process would be scheduled using a final assembly schedule (FAS). The FAS would happen after the receipt of an order.

In a make-to-order (MTO) environment, the SKU driven by the master schedule might be the raw material SKUs or possibly the weldments, depending on the BOM complexity and lead time of the customer requirement versus the accumulative lead time of operations. *Probably* if you went into this shop at

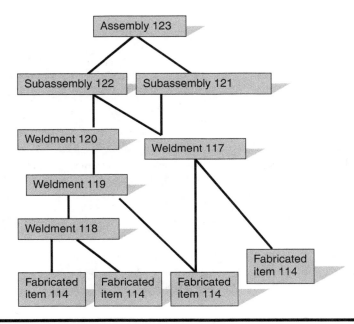

Figure 7.5. BOM Structure for MTS Assembly in Figure 7.4.

any one time, you would find raw material stocked waiting for the customer order, but not necessarily.

Each of these inventory strategies has its costs and benefits. Very few companies have only one strategy to fit all products. Most have several. This is normally the most efficient way to run a manufacturing business, although there are always exceptions to that statement. Commodity products are most often the markets where products are MTS. Recent learning in the lean space has been especially applicable to inventory strategy due to extra inventory cost acknowledgment. Commodity products are often the lowest-margin products in the market, and the MTS strategy tends to have more cost associated with it due to inventory requirements. Lean processes drive flexibility and speed, thus reducing inventory requirements and allowing more SKUs at the finished goods level to be moved to ATO strategy.

CAPACITY PLANNING IN THE MASTER PRODUCTION SCHEDULE

There are two types of capacity in this discussion: demonstrated and theoretical. It is extremely important to distinguish between the two. At The Raymond

Corporation, while I was working in scheduling, there was a welding manager who had been with the business for more than thirty years. I learned not to ask him about capacity because his answer was always influenced by his enthusiasm and memories of the past. Back then, Raymond was mostly MTO, not welding mainframes until after the customer order was received. In those days (before lean), mainframe weldments on newer designs often took more time than on earlier simple mainframes because of the increasing complexity and sophistication of new models. This gentleman would always answer mainframe capacity questions with a slant toward what they *used* to be able to do. The difference between reality and his memory was considerable. I quickly learned to use the facts, not the opinions, even if from experienced and knowledgeable people.

Capacity is a major concern for the master scheduler, probably the most important factor governing the scheduling process. A high-performance master scheduler always uses demonstrated capacity and is not easily fooled by opinions and emotions without backing in data or facts. Early in this chapter, we looked at an MPS about a week out. Figure 7.6 is similar to that illustration, but opens up the question of capacity and good scheduling technique.

In the earlier depiction, "capacity" was shown at level A. In Figure 7.6, there is an imaginary level of capacity, just above the filled bucket value of capacity scheduled for consumption. The inference in the diagram is that the master scheduler should not schedule to the maximum demonstrated capacity all the

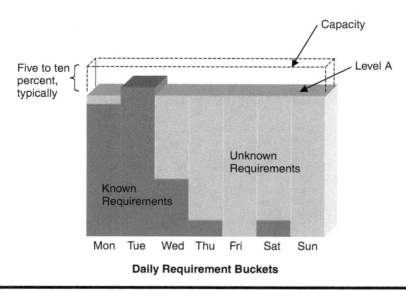

Figure 7.6. Master Schedule Known and Unknown Requirements.

time. To do this is to fail at master scheduling. Demonstrated capacity is equivalent to what the process can generally yield on any given average day. It does not mean that the process will always deliver at that level. Sometimes it will be less, sometimes more. Good master schedulers take their responsibilities very seriously. When they miss schedules, it is a reflection on their accuracy and performance. Class A master scheduling requires that schedules are hit 95+ percent of the time. To accomplish this, demonstrated capacity must be well understood and schedules need to be synchronized with reality continually. At Class A ERP performance levels, this must happen at least once a week. In a Class A ERP environment, synchronizing the schedule is not that difficult and should not be compromised. Any consideration of an alternative practice (not regularly synchronizing the MPS with reality) is a violation of Class A master scheduling performance criteria and should not be done.

While we are on the topic of reality, let's consider customer order entry as a factor in predictability. In most businesses I have worked in, customers are generally ill behaved. Orders do not always come in at the same rate every day. Some days orders are heavy, and other days orders are light. Does this sound familiar? It does to most companies. Let's look at a possible schedule going forward from the schedule illustration in Figure 7.6. In Figure 7.7, let's say that customer orders in this MTO environment actually were entered with unequal

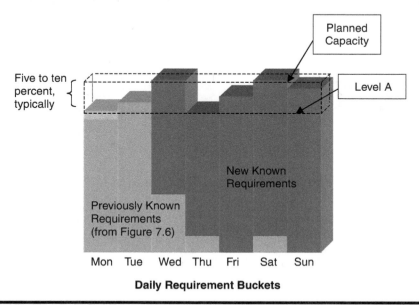

Figure 7.7. Master Schedule Known and Unknown Requirements.

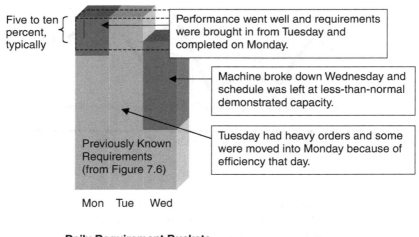

Five to ten percent, typically

Performance went well and requirements were brought in from Tuesday and completed on Monday.

Machine broke down Wednesday and schedule was left at less-than-normal demonstrated capacity.

Previously Known Requirements (from Figure 7.6)

Tuesday had heavy orders and some were moved into Monday because of efficiency that day.

Mon   Tue   Wed

Daily Requirement Buckets

Figure 7.8.   Massaging the Master Schedule Daily.

rates of demand from one day to the next. Figure 7.7 might show the results of an uneven demand. If the original planned schedule was filled to the demonstrated capacity minus 5 percent, there would be some flexibility to move and shift orders without impacting customer promises in a big way. In many high-performance environments, there is a shifting of orders each day depending on the actual orders completed. Figure 7.8 demonstrates this process.

This kind of constant schedule shifting and alignment is necessary for a high-performance master schedule. The root cause is customer variation, a process input that almost all businesses experience, regardless of market or product.

In reality, Figure 7.9 probably reflects how orders can come into your order entry process. No matter where you are in the food chain, consumer spending

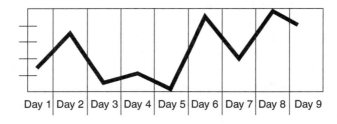

Day 1 | Day 2 | Day 3 | Day 4 | Day 5 | Day 6 | Day 7 | Day 8 | Day 9

Figure 7.9.   Normal Customer Order Fluctuation.

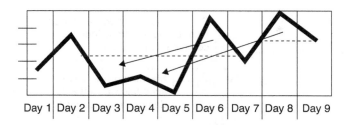

| Day 1 | Day 2 | Day 3 | Day 4 | Day 5 | Day 6 | Day 7 | Day 8 | Day 9 |

Figure 7.10. Smoothing Normal Customer Order Fluctuation.

affects demand, and consumers can be finicky, to say the least. It is probably appropriate to say, "We met the enemy and it is us as consumers!" If Figure 7.9 was a picture of the production schedule, it would not depict how most companies want to run their operation on the supply side. For that reason, some smoothing is done regularly (Figure 7.10). Some is done in the operations planning process, but the MPS is where it is controlled. This is where planned orders are impacted by the customer.

The process that works the best to level load this type of order behavior is to move the spikes into the dips. No rocket science here. When the MPS is not fully loaded to the maximum level of demonstrated capacity all the time, this is easily facilitated. The fears of loading less than the maximum are dispelled quickly when performance improves overall.

A plant manager friend (Scott) recently took over a new position in a large established plant with a lot of history. The plant was having continued difficulty achieving sufficient levels of production. Line managers were talking about capacity restrictions as if they were insurmountable. The first thing Scott did was to instruct the master scheduler to lower the MPS by two hundred units per day, a reasonably significant percentage of current schedules. Keep in mind that this was in an environment where the demonstrated capacity was less than regular customer-driven demand. The human response was predictable. From the first request, the first-line managers within the plant were convinced that this new plant manager was clearly wrong and dangerously fooling with the schedule! Why would he decrease the schedule when the plant was not making enough already? Surely that would limit flexibility because less inventory would be positioned for the daily build; besides, there were always shortages. More inventory gave them more choices each day for sorting out something to build. By now you have the picture. The result of the schedule decrease was also as predictable — the average daily rate of production *increased* immediately by twenty units per day. This understandably may not make sense to those who have not seen the results of poor master scheduling in action. I meet people every day who share the fear that decreasing the MPS will decrease the build performance. This is true only when the MPS is scheduled to proper levels to begin with. Given this environment, it is clear that the inflated schedule was actually slowing the performance down. It was also costing much more in material handling, schedule changes, and inventory capital.

Knowing and understanding the demonstrated capacity and scheduling some small buffer of flexibility into the schedule will provide the best results. If the demonstrated capacity fluctuates up and down more than seven to ten percentage points, day to day, there is great opportunity and a need for a Six Sigma–type project where root cause is analyzed and actions driven to reduce variation. It is difficult to schedule the correct amount if there is so much process variation that demonstrated capacity is difficult to understand or recognize.

> It is difficult to schedule the correct amount if there is so much process variation that demonstrated capacity is difficult to understand or recognize.

Keeping flexibility scheduled from the viewpoint of the MPS almost always will help. The worst that happens is that you move work in from the following day if the schedule is met 100 percent as planned with spare time left in the day. There is an underlying assumption with this statement, by the way, that inventory is brought in at least twenty-four hours prior to need. Most organizations have required dates at least twenty-four hours in advance. There are exceptions to this in the automotive industry and in well-executed lean pull systems. If you fit that mold, an indicator would be shown as more than one hundred turns on inventory per year. I do not run into many organizations with that performance, although there are some out there. If you are in the fifty-plus turns a year, it is almost a given that you are executing to a well-disciplined MPS. Congratulations!

Level loading is accomplished through best understanding the demonstrated capacity and having a reasonable handle on forecasted demand. That is not to say that the forecast has to be accurate all the time or even most of the time. It simply means that the forecasting process does exist and the demand-side folks are playing ball and working to facilitate compromise when necessary. Level loading is best done not through loading to maximum capacity all the time, but to levels just below the demonstrated maximum capacity. Consistency in process is the enabler to increase the MPS and decrease the buffer between schedule and

> Consistency in process is the enabler to increase the MPS and decrease the buffer between schedule and demonstrated capacity maximums.

demonstrated capacity maximums. This may be the most misunderstood principle within master scheduling.

RULES FOR LEVEL LOADING

Level loading at the global level happens in the operations planning, but daily and/or hourly level loading is also necessary in most businesses. The following rules are appropriate after the operations plan monthly level loading has been accomplished:

1. Schedule materials to be available (not received, but available) at *least* one period prior to the required need. If the scheduled time/requirement buckets are in days, allow supplier releases (only as authorized) of material twenty-four hours ahead of requirements. If the schedule is in hour time buckets, schedule availability one hour prior to need. If the schedule is in minutes, schedule the material to be available one minute prior to need. (I have yet to visit a company with the MPS schedule in seconds, but I can visualize it.) This may sound like waste, but if you want to execute to high levels of performance, some material flexibility is required. Most businesses overdo it, however, by bringing material in way before the need. These businesses often perform at less than twelve turns on their inventory.

2. Schedule each product family separately. Among the obvious reasons of shared constraints, this is necessary because inventory strategy affects product family selection, and this strategy dictates the level within the BOM and facilitates proper scheduling.

3. Understand demonstrated capacity of the product family. "Demonstrated" means "normally executed levels," including "normally experienced process variation" (*unusual* process variation is variation not experienced normally).

4. Create weekly schedules that allow for some execution variation. Schedule some buffer time above scheduled completions up to full demonstrated capacity.

5. When performance is robust and no significant or out-of-the-ordinary process variation is experienced, move future requirements into the present period to fill any unused buffer. This can be done automatically from the factory floor if the internal rules of engagement are specific enough to this need.

6. If no orders exist to pull up in the next period or there are no customer orders convertible to cash, allow employees to work on process improve-

ment projects. Some suggestions include 5-S projects (house cleaning and workplace organization) or setup reduction (both machine and assembly areas), preventative maintenance, etc. This is valuable time and must be put to good use.

LINKING TO THE OPERATIONS PLAN

The MPS is the detailed translation and convergence of both top-management planning and actual customer activity. In the beginning of the month, the MPS will be exactly in line with the operations plan, by product family (Figure 7.11). If the customer activity deviates greatly from the anticipated top-management planning process, the MPS needs to "take flight."

In essence, the MPS is always in alignment with the operations plan from the S&OP process at the beginning of the monthly cycle. By the second week of the month, if the operations plan is inaccurate, the MPS goes its own way, correcting the requirement drivers in accordance with actual demand experience. As the Class A metrics will show, the operations plan in that scenario will show performance issues and the MPS will show improved performance as the plan is realigned with actual process performance. It is the master scheduler's responsibility to keep the MPS in line with reality.

Time Fence Rules in Master Scheduling

There are several theories on and approaches to fence rules in scheduling. Class A rules of engagement normally include specifics for each product family documented as time fence rules. Without these rules/handshakes, it is more difficult to have high performance with the optimum efficiency of operations. In a normal process, the lead-time elements might look like Figure 7.12.

Your process may look different than this, but usually procurement exists and at least one other conversion process exists as raw material is converted

Situation	Characteristics of the MPS
When demand equals S&OP plan	MPS follows operations plan through month
When demand exceeds S&OP plan	MPS deviates/corrects plan as required
When demand is less than S&OP plan	MPS deviates/corrects plan as required

Figure 7.11. Class A ERP MPS Alignment with the Operations Plan.

Procurement Components	Fabrication	Subassembly	Assembly	Ship to Customers

Figure 7.12. Accumulative Lead Time.

into saleable product. If your process is simpler, you may have some interpretation to do. In Figure 7.13, the fences are allied to this lead time.

Fixed Fence Rules

The "fixed" fence is sometimes referred to as the frozen fence. No schedules are ever completely frozen. Because of this, most companies have abandoned the term "frozen." In Figure 7.13, the inference is that more flexibility is built into the schedule. At each time fence, the rules change just a little and all the players in the supply chain are aware of the rules and the latest schedule. Flexibility is part of this handshake at each time fence interval.

Normally, the fixed fence is no more than one week. In many businesses with lean processes and flexible suppliers, the fixed fence can be as short as one or two days. The rules are normally simple — no changes. With that type of agreement, there is also a need to define the rules for how to *make changes* within the fixed period. Usually, changes inside this period have some impact on cost, and rules may define who picks up costs to certain quantities or time requirements. When suppliers are making only enough to cover the fixed period in the short term, this can mean overtime or special freight, etc. In the best examples, the customer picks up charges within this fixed fence. After all, the idea within this time fence is to keep the process variation to a minimum. A common scenario has the supplier focused on exactly what the fixed schedule

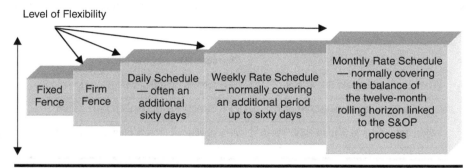

Figure 7.13. Time Fence Norms and Flexibility Built into the Planning Horizon.

calls for and stocking only a couple of units (or components for units) over the demand. This gives some flexibility, but for cost reasons is limited within this short period. A company in Mexico that supplied components to Caterpillar had agreements with its customer that the fixed fence would be two days. Even within that short lead time, quantities would change at times. This supplier had two or three extra assemblies for all of the most popular items available all the time. A pull system kept this buffer stock filled. Caterpillar knew the buffer quantities, and a handshake rule allowed changes up to the buffer level. This type of agreement works quite efficiently. It also does not have to be dictated to your customer. It can be a simple agreement for general practice. The important part is to have the customer aware of this process flexibility while everyone is keeping unplanned costs to a minimum.

Firm Fence Rules

Beyond the fixed fence, more flexibility would be expected in most supply chains. This requires a handshake within the supply chain and might allow up to plus or minus 20 percent on requirements communicated at the beginning of the period. For example, on January 1, in an environment with a three-day fixed fence and an additional seven days in the firm fence, the schedule says one hundred units per day of a specific part number through the entire ten days. The flexibility agreement within the fixed fence may be plus or minus two units per day. On the fourth day out, flexibility requirements are increased to (possibly) plus or minus twenty units per day. These types of handshakes allow everybody to develop the right inventory strategy and keep costs to a minimum. These rules are part of the risk management for the supply chain. It is in the interest of both parties to have the agreements at a sensible level. The wider the schedule swings can be, the more capacity needs to be committed that can drive cost. This is also why it is so important that the schedules be as accurate as possible.

> The wider the schedule swings can be, the more capacity needs to be committed that can drive cost.

Daily Schedule Beyond Firm

It gets easier to define flexibility the farther out you go on the time line for product delivery. Beyond the firm fence, the master scheduler would normally schedule detailed daily requirements out for some reasonable time frame. If the fixed period was three days and the firm fence another seven, the daily schedule fence would start after that, maybe from day eleven to day sixty. This means that from day one to day sixty, all the requirements in the MPS are scheduled in daily buckets with very specific SKU information driving the supply chain. The flexibility rules keep getting more demanding the farther from the current period you go. Many times, the rules of engagement with suppliers can increase flexibility requirements to plus or minus 20 percent out beyond the firm fence.

Weekly Rates Beyond the Daily Schedule

Beyond the daily schedule in the MPS, master schedulers will often migrate to weekly time buckets. Often, the farther you get from the current period, the less accurate the requirements are anyway. Keeping the level of detail within the schedule at the SKU makes a lot of unnecessary work that will probably be changed several times anyway. It is often much more efficient and accurate to use planning bills and insert weekly rates by product family for the next few weeks. Flexibility agreements are often wide open at this point and are a matter of negotiation since the period is often outside accumulative lead time for the supplier anyway. It is normally in the best interest of all parties to increase as required, especially when there is time to prepare for the increase.

Monthly Rates Beyond Weekly Rates

For the same reasons, master schedulers move to *weekly* rates at some point within the twelve-month rolling horizon; out at some point, the level of detail can be minimized again. This is done by simply moving from weekly to monthly rates by product family. This keeps the maintenance work to a minimum and allows for easy alignment with the S&OP process monthly. Again, flexibility requirements are the main driver for capability requirements and schedule simplicity.

The preceding rules are defined in this text as a general method. Your situation may vary, but the concepts as described are valid in every business. Obviously, when suppliers are on the other end of the globe or critical components have long lead times to manufacture, it becomes even more important to define these rules of engagement with the suppliers.

> When suppliers are on the other end of the globe or critical components have long lead times to manufacture, it becomes even more important to define these rules of engagement with the suppliers.

Each supplier can be different and may have specific rules, but the general rules should be applied whenever possible to keep things simple. Make sure they are written down and communicated. In a Class A ERP environment, these rules are followed. The master scheduler does not always negotiate these rules with the suppliers, but certainly is part of the team.

MASTER SCHEDULER'S ROLE

There are several important roles the master scheduler plays in a manufacturing organization. Duties start with the monthly cycle of updating the operations plan for top management, and the master scheduler plays a vital role in the S&OP process. Although the operations plan is blessed by management, it is rare that top management actually develops the plan.

> Although the operations plan is blessed by management, it is rare that top management actually develops the plan.

Following the monthly S&OP process, the master scheduler is responsible for developing the detailed plans for production to follow. In the middle of the week, usually on Tuesday at 1 P.M., the master scheduler often facilitates the weekly performance review meeting, the accountability infrastructure for Class A ERP performance. His or her activities also include communicating regularly with the order entry function as well as demand-side personnel who are in need of special assistance outside the normal rules. Table 7.1 describes most of the roles and duties of the master scheduler.

The master scheduler role is a big one, indeed. The duties can be summarized into a few categories: communication between sales and operations, develop the operations and master schedules, responsible for process linkage for schedules, coordination of scheduled capacities, maintenance of the twelve-month rolling capacity buckets, and development and enforcement of the rules of engagement. This heavy load requires that the position be filled with a capable person. Generally, this means product knowledge, leadership capabilities, and materials and operations experience. One of the best master schedulers I ever met was actually the maintenance manager before going into sales management and then moving to MPS manager. The following list depicts backgrounds of people I have worked with who made great master schedulers:

- Materials manager
- Demand manager
- Maintenance manager
- Production manager
- Product planner
- Sales manager
- Industrial engineer

As you can see, the backgrounds of this group are so varied that background alone may not be that important. What is important is the ability to command

Table 7.1.   Roles and Duties of the Master Scheduler

Role/Duty	Explanation
Develop the operations plan	Using up-to-date information, propose operations plan alternatives
Update the president/CEO for S&OP	Meet beforehand to ensure specific questions get asked in the S&OP
Facilitate the weekly performance review	Is master of ceremonies for this important performance review
Massage the weekly schedule	As orders are received, schedule accordingly
Attend the weekly demand review	The master scheduler is a main player in the weekly demand review
Process owner for MPS metrics	Complete root cause analysis and drive actions to eliminate variation
Develop rules of engagement	Develop and maintain rules of engagement with the demand side
Coordinate schedule changes	Update and change schedules as required on a daily basis
Ensure MPS content	Responsible for all business requirements to be driven in the MPS
Chair the weekly clear to build	Facilitate the production and procurement buy-in of the weekly schedule
Determine planning BOM design	Using knowledge and data, develop planning BOM structure
Team member for new product introduction	New product introduction is a major factor in MPS accuracy
Involved in engineering change	Effectivity dates of changes are often coordinated by the master scheduler
Master schedule maintenance	Create the bucket size based on demand forecasts and S&OP
Communicate lead times to the field	If lead times fluctuate due to inventory strategy, communicate to sales
Liaison between sales and operations	Facilitate proper communication between sales and operations
Work with demand manager	Regular communication with the demand side

respect and lead people. Product knowledge is also important, as is knowledge of the process idiosyncrasies of the firm. Also notice that the prerequisites for this job are generally of a management nature. This type of background helps with communication skills. The master scheduling responsibility is truly a management position. Many master schedulers have other functions reporting to them. Figure 7.14 shows a typical yet simple master scheduling organiza-

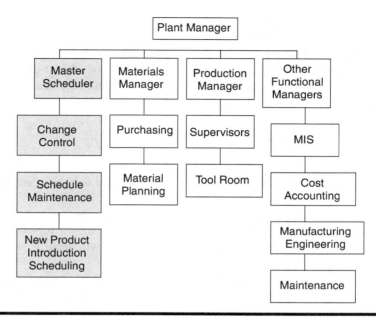

Figure 7.14. Typical Master Scheduler Reporting Structure.

tional reporting structure within a multimillion-dollar multiplant operation/ environment.

MASTER SCHEDULING METRICS

Any Inventory Strategy with a Minimum One-Week Handshake

The master schedule is normally a weekly process within the twelve-month rolling plan inside the ERP business model. Accordingly, the metric should be a weekly view. In most high-performance organizations, the master schedule is locked (for measurement purposes only) at the end of each week. Most of the time, this is either Friday or Saturday, although that is not important as long as the process is consistent, week to week. When the schedule is locked for measurement, the plan becomes the plan of record. Often, the schedule is drafted by Thursday night and production and procurement management are asked to give their blessing on Friday, which means they agree to be held accountable for it. That blessing happens in what Class A ERP refers to as the clear-to-build (CTB) process. The CTB is a component of the master scheduling management systems requirements.

The metric is driven from the locked plan described above. The plan of record is reviewed at the end of the week to see how close the original weekly plan was. The rules surrounding the metric follow. This may seem like a lot of rules, but there are always questions and this list should answer most of them.

Rules for Measuring Accuracy

1. The metric is the percent of complete orders that are completed in the week they were scheduled according to the latest weekly schedule (plan of record from the end of last week).
2. There is a separate schedule (and metric) for each product line.
3. Each schedule is locked at the end of each week for the following week (for measurement purposes only; the schedule will actually be revised as necessary).
4. The plant metric (consolidated) is the accumulated percentage of orders completed within the week scheduled. No averaging of averages required here.
5. If there are some product lines with numerous small orders and others with few large orders, it does not make any difference! (See rule 4 above.)
6. The planned schedule is to the detailed configuration (SKU) and order quantity.
7. If the orders scheduled for completion are all completed with full quantities, the plan is met and the measurement is 100 percent.
8. Any orders that are not completed to full planned quantities are misses in the metric.
9. Product that does not make it through normal quality checks in the assembly process by the end of the week and creates missed orders also creates misses in the master scheduling metric.
10. Products that need rework, if reworked within the scheduled time, are schedule hits, not misses. Note that it is not the intention of any process to require "efficient rework," but other measurements will pick up the first-time yield opportunities. The master scheduling metric simply is designed to focus on schedule accuracy.

The metric is calculated just as any percentage. The numerator is the number of orders completed to the original plan of record compared to the denominator, which is the number of orders planned:

$$\frac{\text{Number of completed orders (right quantity)}}{\substack{\text{Number of orders planned} \\ \text{in most recent plan of record}}} = \text{Percent MPS accuracy}$$

Remember that it takes two processes to happen for Class A ERP performance criteria to be met: a good plan and good execution. When both happen, it is a good day at the office!

Any Inventory Strategy with Less Than One-Week Handshake

There is a little twist to the metric when building short-cycle products, especially in an ATO, MTO, or ETO environment. In these environments, the plan of record when locked on Friday the prior week can be mostly unscheduled orders. Take, for example, a saw blade company that receives orders for big saw blades. This company makes blades up to forty-two inches that are made per customer application for cutting hard materials like brick, concrete, and even metal. The company makes the blades specifically to the application from scratch and ships in less than seventy-two hours. In this organization, the master scheduler has no idea what will be built and shipped two days from now, to say nothing about what will happen at the end of the week. The master schedule consists of planned capacity to be held, and as orders are received, this capacity plan is consumed. The metric is executed a little differently in this environment. The metric is still compared, actual to plan of record, but the plan of record is the first time the order is scheduled. All orders that are completed in the week they were originally scheduled are considered hits (not misses).

To some people, this metric may seem easy and too forgiving. There is no need to be concerned after the entire Class A metrics package is put together. The master scheduling metric makes sure that the weekly master schedule is as accurate as possible. Class A requires a minimum of 95 percent performance overall in this calculation. Again, it takes a good plan and good execution to make a Class A process. A good master schedule linked to the S&OP process is a great start!

MANAGEMENT SYSTEMS REQUIREMENTS

"Management system" is a term often associated with Class A ERP. The term is used to describe an autopilot infrastructure for management follow-up. There are several management system events required in the Class A ERP process.

Class A ERP requirements dictate the necessity of system maintenance. At least once a week, the system needs to be synchronized with reality completely. If some orders in the previous week were missed or skipped for whatever reason, the following week needs to start out with complete realignment of schedules. The same is true if work was done ahead of schedule in the previous week. It is not acceptable, desirable, or logical to have known inaccurate requirements at the start of a weekly master schedule cycle. Past-due requirements are especially harmful. Since there is no such thing as accuracy in past-due schedules, there should never be any "past dues" in the master schedule, especially as the weekly schedule begins.

> Since there is no such thing as accuracy in past-due schedules, there should never be any "past dues" in the master schedule, especially as the weekly schedule begins.

The management system element in this space is a weekly commitment to maintenance. This needs to be done by the end of the week — always — without exception. If the master schedule is kept up to date regularly, and schedules are truly accurate each week within 95 percent, the maintenance process will take almost no time.

The weekly CTB is the next management system Class A ERP requirement. The deliverable from this process is a handshake between master scheduling and the two critical support areas: production management and procurement managers. This handshake facilitates accountability, a prerequisite for Class A ERP performance. The CTB event does not have to be a meeting. It can be as simple as an e-mailed schedule sent to all production managers and procurement on Thursday afternoon with an agreement that the approval is assumed if there is no reply by Friday at 1 P.M. In one bakery organization I worked with, the master scheduler in Baltimore faxed the schedule each Friday morning to the plant managers located in bakeries in various cities around the United States. The plant managers had until 2 P.M. that afternoon to respond. If there was no response, by documented agreement, the handshake was assumed.

In other companies, the preferred method is to have an actual review meeting Friday morning to go over the schedule with all of the players. In these environments, the plant manager is normally also present. The only thing that dictates the specifics of how to run this event is management backing and production, procurement, and master scheduling trust. The more trust and accuracy there is in the process, the less formal the event needs to be. What-

> The more trust and accuracy there is in the process, the less formal the event needs to be.

ever the case, Class A criteria require this handshake. Process ownership, no matter how trustworthy the parties are, requires a chance to say no.

The pinnacle of process ownership is experienced at the Class A weekly performance review. This meeting is normally facilitated by the master scheduler and, except for the S&OP process, is probably the most important meeting in the Class A ERP criteria. The weekly performance review is an event where the process owners from all aspects of the Class A process can show what they have done in the last week to improve their process performance. These process owners are required to report in a standardized format with performance numbers, trend analysis, root cause, and actions/names/dates. By having the master scheduler run this meeting, the plant manager is left to listen, praise, and ask the tough questions. Having the master scheduler run this meeting may seem out of his or her responsibilities, and if that is what you are thinking, you have not yet accepted that the master scheduler is the second highest-ranking position behind the plant manager in most high-performance facilities. When you put it into that perspective, the role in this weekly performance review is understood more easily.

> Having the master scheduler run this meeting may seem out of his or her responsibilities, and if that is what you are thinking, you have not yet accepted that the master scheduler is the second highest-ranking position behind the plant manager in most high-performance facilities.

PROCESS OWNERSHIP

Process ownership in the master scheduling environment is making sure that the schedules are in sync with the S&OP, are accurate, and are communicated properly. In addition, it means having a twelve-month rolling horizon in the business system for the supply chain to see and maximize efficiencies. The schedules have to be updated/reviewed at least once a week in a Class A environment and accuracy must be at 95 percent.

The process owner is obviously the master scheduler for this process. He or she will document the performance and opportunities and report to the weekly performance review to review accomplishments and progress with the operations team. The master scheduler also facilitates the weekly CTB and works with top management to both prepare them and answer questions prior to the S&OP. During the S&OP, the master scheduler is a messenger and is responsible for the accuracy of data, but is not necessarily the one who has to answer the tough questions about missed performance. The VP of operations, VP of sales, and/or CFO answer the questions in that venue.

As in any process, there needs to be a succession plan. For any of the responsibilities discussed, when the master scheduler cannot attend an event or meeting, a backup should be trained and authorized to act with full decision-making responsibility. This should go without saying, but to meet Class A certification criteria, it must be so.

The master schedule is the driver for all production activity including the supply chain. It is a critical process and one of the most important in any business. The handoff from the MPS is the material planning functions within the organizational flow chart. In the next chapter, material requirements planning will be discussed.

8

CLASS A ERP
MATERIAL PLANNING

At the heart of almost every material planning process within manufacturing is some form of material requirements planning (MRP). MRP has gotten a bad reputation in recent years. Among the "constraint" advocates, the Six Sigma people, lean gurus, software visionaries, and the just plain inexperienced types, the word on the street is that MRP is dead. Nothing could be farther from the truth. When the discussion point is strictly described as "netting of future requirements," few would argue about the need to do this in manufacturing planning both now and in the future. The truth is, almost every major successful manufacturing business on the face of this planet uses some form of MRP, whether they want to admit it or not.

The software organizations have as much to do with this confusion as anybody, and it is understandable. Much like the label "station wagon" is out of date (the term is now SUV), MRP also has a few new names. Most of these names are recognizable by hints like "advanced scheduler," "advanced planning and scheduling module," or even "drop and drag order sequencer." These newer labels are the "four-wheel-drive SUV" versions of the MRP station wagon! Some of the names are actually quite clever. At the end of the day, however, these planning engines are basically netting requirements and time phasing the resulting signals for either procurement or manufacturing. The gains are in automation and ease of use, just like automotive technology has given us outside temperature readings, GPS, and a digital compass in the new "station wagons."

Material planning is the process of taking the requirements handed off from the master production schedule (MPS) and determining what, if any, compo-

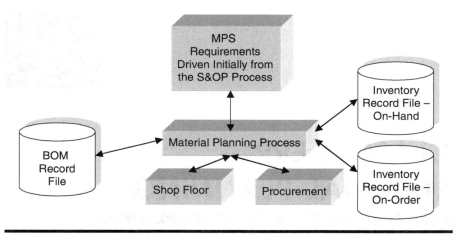

Figure 8.1. MRP Inputs and Outputs.

nents need to be ordered versus those already available. Availability is derived from stock, being in the process of procurement, or being manufactured (see Figure 8.1).

The MPS requirements are the real drivers of material planning. As described in Chapter 7, the master schedule in most high-performance environments is massaged frequently to synchronize the supply side with customer demand. The material planning engine in the ERP software business system is the calculator that determines what to do with these MPS signals through requirements netting. For the rest of this discussion, MRP will be the label referred to when describing this detail-planning engine.

MRP has several activities to perform once the MPS requirements are refreshed. The first activity is to determine requirements from the bill of material (BOM) record for each requirement. The BOM record holds or is linked to several pieces of important information. These data include:

1. Top-level part number (parent-level number)
2. Component part numbers called out in the upper level
3. Usage per component in the upper-level assembly
4. Unit of measure for each component
5. Lead time and lead-time offsets for each component

Once the BOM information has been accessed, the next information the MRP engine requires to complete its job completely is inventory records. There are two types of inventory records for this discussion: on hand and on order.

The names are descriptive enough, but for clarity there is an important Class A influence on accuracy in this space. The importance of data accuracy becomes critical as realization of data dependency is underlined. For this planning engine to do its job effectively, the BOM records must be impeccable and inventory records must be pristine. Class A ERP criteria require BOM accuracy to be at 98 percent to achieve a minimally acceptable level and 99 percent to be considered good. Inventory record accuracy needs to be 95 percent to meet the threshold of acceptability and 98 percent to be considered robust. These are important requirements to keep process variation and associated costs to a minimum. In both cases, the performance reflects the percentage of perfect records, not how accurate each record is.

If the master schedule is at a 90 percent level of accuracy in either quantity or date, MRP will be affected greatly. If the process variation exposure is increased by having less than 98 percent BOM accuracy or if the inventory record accuracy is below 95 percent, the accumulation of process input variation becomes an almost *exponential* problem affecting the output accuracy of MRP. This is not a complicated idea. It is simple logic. It should be the goal of any organization to use the business system on autopilot in every opportunity possible to lower cost. Bad data eliminate autopilot as a possibility. Human intervention is immediately and consistently required to offset process variation caused by inaccuracies in

> Too many organizations today still do not appreciate the huge payback available from such simple efforts as process transaction disciplines.

data. Too many organizations today still do not appreciate the huge payback available from such simple efforts as process transaction disciplines. More discussion will be provided in Chapter 9 on data accuracy elements.

ORDER POLICY DECISIONS

MRP generates signals according to the calculation resulting from inputs and outputs that the planners need to respond to. When a requirement is determined for a particular component, one of the first questions the planner has to answer is how many to order. Sometimes the decision is easy (for example, only the number to match requirements), but other times it is not as simple. Consider a situation, for example, where a relatively inexpensive item is purchased from a supplier located thousands of miles away. If the requirement is for stainless steel bolts at a quantity of seventeen with a price of thirty cents and the source for these bolts is Asia, how many is the right number to purchase? The answer, even given the data of price and source, is still "it depends." It depends on inventory strategy of the item and the upper-level requirements, and it depends on anticipated usage beyond any known history. Order policy begins to address the issues and answers the question "How many should we order?" The more popular order policies are:

1. **One for one** — This is the just-in-time, lean approach. No extras are ordered; nothing is ordered until there is a customer demand that creates a requirement. If seventeen are needed, only seventeen will be ordered. This approach is used in an engineer-to-order (ETO) or make-to-order (MTO) inventory strategy. In a lean environment, finished goods are often made on either ETO or MTO strategies.
2. **Lot for lot or order for order** — This policy is a deviation on the approach above. In this situation, subassemblies may be requisitioned in anticipation of a customer order and require components. Items with this order policy assigned are not ordered unless there is a requirement from another parent level. In that case, the quantities are requisitioned to match the upper-level lot size.
3. **Time-value or fixed-period order policy** — With this policy, parts are ordered in quantities to cover the usage for a certain period of time, often a week or more. If the assigned time value for a part is two-weeks' worth and the usage is expected at three per day for two five-day weeks, the order size would be thirty pieces (two weeks times five days times three per day).
4. **Fixed order quantity** — This policy dictates certain predetermined quantities. In some process flow businesses, these quantities are also referred to as campaigns. There can be many reasons for this practice, such as how many can be made from a sheet, how big the vat is in the process, what the normal demand is for the item, etc.
5. **Economic order quantity (EOQ)** — Very few organizations use EOQ any more. The formula is the square root of 2AS over IC, where A = cost of an order to process, S = units per year, I = inventory carrying charge, and C = unit cost. Today, most experts realize that EOQ is not really valid in practice; it is valid only in theory. The main problem with EOQ is the ability to adjust all the variables and end up with whatever lot size you wanted anyway. Whenever I encounter an organization that says it utilizes EOQ, a red flag goes up in my mind.

> The main problem with EOQ is the ability to adjust all the variables and end up with whatever lot size you wanted anyway.

ABC STRATIFICATION

Along with order policy code, stratification also plays an important role in ordering inventory from MRP signals. Inventory stratification is the segmentation of items into layers by monetary or usage values. These layers are known as A, B, and C. Occasionally in some businesses, there is a fourth layer, and yes, it is called D. Each layer has a specific formula for defining the criteria.

In the early 1900s, Italian economist Vilfredo Pareto developed an equation that described unequal distribution of wealth in Italy. He observed that 20 percent of the people enjoyed 80 percent of the wealth. Later this formula became known as the 80/20 rule, and it has proven useful in many problem-solving applications. Vilfredo Pareto's 80/20 rule works well when administering traditional ABC methodology assignments. Generally, the dividing layers are done by:

A = 75 percent of the monetary value of inventory,
     10 percent of the item numbers

B = 15 percent of the monetary value of inventory,
     10 percent of the item numbers

C = 10 percent of the monetary value of inventory,
     80 percent of the part numbers.

As stated, some organizations also have a D category that divides the C items into two groups. D items are usually components that are *very* inexpensive (such as common small washers, screws, etc.), with values much less than one cent each).

Most ERP business systems can automatically calculate the inventory layers and give the user parameters that can be set as desired. Stratification is done to make sure there is more emphasis on the items that have the most risk associated with them. Therefore, critical components that are difficult to get or are on allocation in the market are often forced into the A item code regardless of the ABC calculation.

PROCESS OWNERSHIP IN MATERIAL PLANNING

Normally, in a Class A ERP organization, the materials manager owns the metrics in this space. Class A materials control includes data accuracy, good schedule maintenance, keeping purchase orders and work orders in synchronization with the MPS, accurate management of obsolete and excessive inventory, and keeping data fields in the item master up to date. Lead time is a good indicator of performance in the materials world. Lead time is the critical field that drives inventory decisions and reaction time. It needs to be accurate and as short as possible, but still realistic.

MATERIAL PLANNING METRICS

Metrics in the material planning space are very logical. Class A criteria for planning accuracy allow for some choices here, such as the following suggestions:

1. Percent of purchase orders let without full lead time as defined in the lead-time field on the item master
2. Percent of completely pickable assemblies released on time to assembly
3. Percent of schedule changes within the fixed-period fence (usually forty-eight hours, but can be up to one week)
4. Inventory turns — raw, work in process (WIP), and finished goods.

Each one of these metrics brings some special benefit to the table. Every specific organization needs to determine what the biggest need is in terms of the required drivers of action. If your business is an assembly shop and currently it is common to have orders released for pick without all the components available, the percent pickable metric is very valuable. If your business has myriad changes within the fixed period that cause a lot of cost from changeovers and material movement, then the schedule change metric is recommended. Class A requires two metrics in this area of focus and gives some leeway depending on the circumstances. There are no bad metrics in the list, and implementing all of them can be a risk-free compromise.

INTEGRATION WITH LEAN AND SIX SIGMA

This space is where the most controversy seems to fester around the appropriateness of yesterday's system lessons as they integrate with lean and Six Sigma. Let's get Six Sigma out of the way first since it is justified so easily and logically.

Six Sigma is a process focus that formalizes good problem-solving techniques and establishes data mining and use of facts as prerequisites to success.

Success is defined by the voice of the customer. In this case, we could define a couple of customers. The first customer is the obvious one, the one that pays the bills through product or service purchases. The second is top management and the stakeholders of the business that ask for return on their investments. Having a lot of material all the time is great for the end-item customers, but the stakeholders would rather have cash earning a return as opposed to having it tied up in slow-moving inventory. Using statistical tools from the Six Sigma kit does nothing but help the application of MRP. Here are some examples of Six Sigma projects in companies that have paid back many times and connect directly to MRP and materials planning:

1. Increasing forecast accuracy
2. Increasing accuracy of BOMs
3. Decreasing the lead time of major components
4. Increasing throughput yield
5. Decreasing scrap
6. Decreasing dependence on informal systems in the shop

There really is an unending list of possible Six Sigma projects that could help MRP work more effectively. Lean is probably a little more involved, but not any less helpful and easily integrated. Lean is a strategy that focuses on waste elimination. In the early days of MRP, the (then new) calculation actually eliminated some waste, but in today's standards, there still was a lot of waste in the process and the low-hanging fruit was plentiful. This waste was found in the form of excess inventory, large lot sizes, too much make-to-stock (MTS) inventory strategy, and little concern for data integrity. All of these topics have seen major gains in the modern business environment. The "pull system" aspect of lean methodology is a good place to start with the integration discussion.

MRP has traditionally been a "push-type" system, meaning that orders would be made ahead of time in anticipation of customer demand. Sometimes this was at the lower levels and not necessarily finished goods, but too much inventory could be easily generated just the same. If this was the case all the time, it would be easy to kill and bury MRP. There is more to the story than that simple explanation, however. The simple-minded, traditional view is explained in Table 8.1.

In Table 8.2, the vision is upgraded to modern thinking and use of MRP tools. One key element in modern thinking is the range of MRP use. If the planning horizon is divided into time fences (fixed, firm, and planned), the rules and techniques are clearly different in the various time fences. A hybrid MRP methodology has developed to help utilize the strengths of both MRP planning

Table 8.1. Traditional Thinking: MRP Versus Lean

Topic	MRP	Lean
Inventory strategy	MTS	MTO
Inventory level	High	Low
WIP level	Moderate to high	One-piece lots
Rate	To level load	To customer demand
Production control	Critical ratio calculation	Visible, kanban
Production communication	Shop work order	Kanban pull
Lot size	EOQ	Customer order or one piece
Lead time	Long (queue, move time)	Short
Focus	Next operation	Customer requirement

Table 8.2. Modern Thinking: Hybrid MRP/Lean Model

Topic	MRP/Lean Hybrid
Inventory strategy	Appropriate inventory strategy with customer draw
Inventory level	Moderate to low
WIP level	Low
Rate	To customer demand
Production control	Visible, kanban, MPS
Production communication	Kanban pull
Lot size	Customer order or one piece
Lead time	Short
Focus	Customer requirement

processes and lean techniques. The result can be very high performance. Figure 8.2 helps visualize the difference.

When MRP and lean are integrated, the possibilities of powerful supply chain management and lean efficient customer draw on inventory and process support Class A ERP and the resulting high performance.

RUNNING THE NET-CHANGE CALCULATION

Since MRP was invented about thirty years ago, much progress has been made in both software efficiency and hardware capacity. The rule of thumb in the beginning of MRP application was to run the regenerative net-change program and calculation every weekend. That was mostly because the program took hours to run and often nothing else could be processed while the room-size computer churned the data, layer by layer. Even more interesting was the

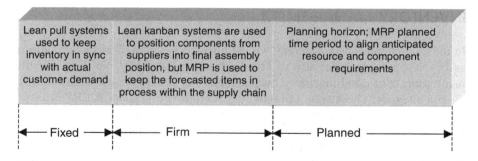

Figure 8.2. Rules in Various Time Fences Using Hybrid MRP Methodology.

philosophy that running it too often introduced a large amount of unnecessary system noise (variation) into the manufacturing process. It is actually quite humorous to think about that today. That system noise was driven by changes in customer demand and data inaccuracies, and the reason it was not-so-affectionately called "noise" was because the manufacturing managers did not like having to change machinery over in the shop. They wanted to run long lot orders. If MRP was not run too often, changes still happened, but they were not communicated to the shop, which artificially created stability in the schedule. This luxury was offset by huge amounts of inventory in finished goods. Expediting was often funded willingly. In those days, heroes were created in the shop and in purchasing daily, and companies like Federal Express became household words. Today, things are very different — certainly in Class A businesses.

> Expediting was often funded willingly. In those days, heroes were created in the shop and in purchasing daily, and companies like Federal Express became household words.

Customer demand is why organizations are in business. To ignore it is foolish and costly. If the existing production process is not matched to the present customer behavior and demand, it is time to re-evaluate the manufacturing strategy. That would include inventory strategy and market offerings. When these strategies are in line with market need, how often you run the MRP program is a nonissue. The more it is run, the more accurate the current schedule is, and life is good.

Most high-performance organizations run MRP at least daily and many run it several times a day. Again, if the inventory strategies and rules governing the supply chain are adequate for the markets you have chosen to serve, there will be nothing but gains from more frequent MRP runs.

INVENTORY MANAGEMENT

In the next chapter, data accuracy will be covered in detail. Transaction design and cycle count processes also will be discussed in Chapter 9, but there are still many aspects of good inventory management that should be addressed outside of data integrity. These areas fall into the materials space. Class A ERP is about predictable and repeatable processes. That includes checks and balances on manufacturing and scheduling processes to minimize unnecessary inventory. Nonetheless, inventory still exists in almost all manufacturing companies. Inventory is the result of a delta or difference between a company's cumulative lead time and the lead time of market requirements. If you have a sixteen-week cumulative lead time, but customers only want to wait three days, there is inventory in the system somewhere, either at the manufacturer or within the supply chain. That is just a fact.

Managing inventory requires a conscience even when the inventory manager does not directly control the ordering. Good warehouse managers in Class A environments have monthly routines that become a conscience for the inventory planning and forecasting people in the business. This role of conscience is sometimes overlooked due to the other normal demands of managing a warehouse. These sometimes overlooked warehouse manager routines include the following list of activities:

1. Run a report of all obsolete inventory items each month. Investigate root cause for this obsolescence. Drive change in the business to eliminate these causes. Each month, there should be a routine of expensing and removing *some* excessive inventory. Communicate a metric of causes to appropriate management within the business.
2. Run a report that includes every inventoried item in the warehouse. Sort this report by current days on hand. Include monetary unit value, total value, last transaction date, and ABC code in the report for each item. Items with more than a few days on hand should be suspect, especially if these items are B or A items. Investigate root cause. In many instances, a Six Sigma project team makes a good vehicle to both gather data and propose a plan of attack for using up these items.
3. Evaluate the preventative maintenance schedule attainment for the period at the end of each month. Make sure the schedules are maintained and are kept current for warehouse equipment.
4. Track changes in volumes and material movement to make sure the items that have the most transactions are the ones closest to the door or docks within the warehouse.

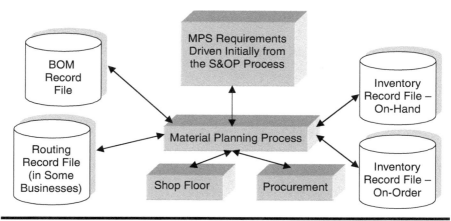

Figure 8.3. MRP Inputs and Outputs.

There are many more duties of a warehouse or inventory manager, but these represent some of the ones most often neglected. In a Class A organization, they are not overlooked. At this point, the focus needs to shift to the inputs of the material planning process and, accordingly, the data integrity of those inputs, represented by the white files in Figure 8.3. In Chapter 9, the focus will be on Class A requirements as well as how to achieve them.

9

CLASS A ERP
DATA ACCURACY
REQUIREMENTS

Data accuracy is an enabler for the enterprise resource planning (ERP) system to work properly and achieve planned results. Class A ERP has very high standards for data excellence. These data elements include inventory record accuracy, bill of material accuracy, item master inputs such as lead time and standards, and — depending on the system and application — routing records. Probably the easiest and yet least robust in many businesses is inventory balance record accuracy.

INVENTORY LOCATION BALANCE ACCURACY

When organizations discuss inventory balances, the usual reference point is the overall total count of on-hand inventory for a certain item number. In Class A ERP, the reference point is a bit more involved. In Class A, the definition of inventory accuracy includes not only the counts, but also the locations of those counts. If there is a total count of one hundred of a certain item "ABC123" and there are thirty pieces in location A, thirty pieces in location B, and forty pieces in location C, the accuracy is in knowing exactly how many are in each location. In many applications, inventory in one location, while usable in a pinch, is not used very efficiently because of the need to move it first, maybe to another line. In yet other businesses, the picking sequence within the warehouse software is only efficient when the locations are known to the system and balances are accurate. This cannot be achieved without good data integrity at the location

Be reminded that it is no more costly to do inventory transactions correctly than it is to do them wrong, and when done correctly, the benefits are many.

balance level. Be reminded that it is no more costly to do inventory transactions correctly than it is to do them wrong, and when done correctly, the benefits are many.

INVENTORY AS AN ASSET

Inventory is a dirty word in most organizations. Once you have "been to the seminar" and/or "read the book," you quickly become a disciple of the **no inventory** program. That is a fallacy in many manufacturing and service organizations. Inventory in the majority of manufacturing or service companies is positioned because of the need to offset variation in their processes. Some of the sources for variation include:

- Process yield rates
- Scrap reporting errors or omissions
- Inconsistencies in amounts produced as compared to requested
- Demand forecast inaccuracies
- Normally ill-behaved customers
- Machine downtime

Understanding this concept is the first step in the ability to eliminate the extra burdens. Having inventory on hand is a great thing when you have a customer ready to buy it. Keeping it around for a few days while waiting for that customer is not so enjoyable.

ALLOWING THE ELIMINATION OF BUFFER INVENTORY

Lower inventory is often a priority with top management in many businesses. When the business goal of lower inventory periodically is brought to the top of the priority list, the materials people get excited about the cause and, not surprisingly, inventory goes down. When lowering inventory, there is only one area in which planners can really help quickly: they can influence levels of new material they order. This same inventory is (hopefully) the component inventory dictated by customer requirements. When an inventory reduction focus is forced on the planners, often the result is increased shortages with no corresponding decrease in process variation. The organization still has the wrong "stuff" inflating inventory monetary values. The shortage and expedite mode that follows translates into self-induced burden cost. The root cause of the problem is often not the buffer itself but the inaccuracy leading to the buffers.

Companies frequently reconcile inventory through physical inventory counts or periodic or frequent cycle counts. Unfortunately, too many of these organizations are really engaged in just "fixing bad balances." Like perpetual cleansing, the counts are designed to clean up ongoing errors that happen unacknowledged. This is not the case in high-performance Class A ERP operations. In Class A businesses, the cycle counts are designed more to understand ongoing transaction accuracy habits than to make corrections.

Let's assume that a business has a posted accuracy of 95 percent. In this imaginary business, there are cycle counts that work through the entire warehouse completely, from one end to the other, every month. If a business assumes that cycle counts are relatively accurate and that the inaccuracies are fixed as found within the four-week cycle, with a posted accuracy of 95 percent, one can assume that at least a 5 percent variation factor is added each and every month. If cycle counts were stopped, the accuracy would fall logically to 90 percent in two months and 85 percent in three. When you look at it this way, it is much less satisfying to brag about a 95 percent performance level. The cost is also significant when that many counts are required every month.

Understanding and eliminating the reasons for inventory inaccuracy is a productive use of resource and eventually results in much less expense to maintain accuracy. One of the easiest ways and most successful methods to uncover the barriers to consistent accuracy of inventory records (determining root cause) is to implement *control groups* in preparation for cycle counting.

CONTROL GROUPS

An inventory accuracy control group is a select group of fast-moving parts or component materials that are repeatedly counted to determine, reconcile, and eliminate the causes of inventory record inaccuracy. It is important to select items/materials for the control group that have high activity. This ensures exposure to error-causing events and/or inaccurate process inputs. The control group should also be representative of the total population of parts/materials. The steps to accomplish this are as follow.

Step 1: Choosing the Items

To start an effective control group, choose thirty to fifty item number locations with high activity. These materials should be a reasonable representation or sampling of the kinds of materials or parts used in the process. It also makes sense to pick stockkeeping units (SKUs) from various areas rather than all from the same area and they should be higher-volume parts to make sure there is activity during the control group period.

The ultimate objective of this control group methodology is to ensure a controlled process and account accurately for inventory movement so that variation causes can be determined. Individual item location balances (rather than the total item balance of all locations) will make up the thirty to fifty balances required for this step.

If there is a question as to how many locations should be counted in the control group, the answer is simple. There is a mandatory requirement to reconcile the variation to root cause each day for all inaccurate balances found. If there are too many inaccuracies in the control group to keep up with and you are not able to reconcile fully each day, you have too many location balances being counted. On the other hand, if you do not find enough location balance variation with the number of locations being counted, you may want to add more locations. This will begin to make more sense as the full control group methodology is explained.

Day One of the Control Group

On the morning of day one, count all of the items/locations in the control group. Next, match the count quantity to the perpetual record file for that location balance (on the computer) and reconcile any differences. The objective is to be assured that the computer record is accurate as the control group process is started. It is important to have an accurate starting point for the implementation of the control group. If any of the balances do not correspond with the perpetual computer record, count them again to ensure accuracy in the count. Make sure that the records you have are absolutely correct. Once the confirmed accurate balances are documented, you are prepared for the next step.

Step 2: Recounting the Locations

On day two, you will repeat the exact same process. Count the same control group of item location balances and match the quantity counted with the perpetual inventory records just as you did on day one. The purpose of this second count is to see if the process variation has shown its ugly head. Often, there is variation occurring regularly, and if you have chosen high-activity locations for the exercise, finding variation will not be difficult. For example, if four location balances in the control group counted have become inaccurate since the previous day's count, some very important information is known at this point, and it is at this point that the value of this process becomes apparent. In this case and all others, you know:

1. These four inventory balances had activity. If you picked the correct location balances for this test, you should be confident that there was activity in each of these locations during the test.
2. Transactions were not aligned with physical movement of inventory. The simple fact that four of the balances in the example were out of line is an indicator that transactions, for some reason, were not in sync with actual activity.
3. The most important piece of information, however, is the simplest. *The error happened in the last twenty-four hours.* Too many times, it is impossible to know when the error occurred, which makes it very difficult to diagnose.

It is critically important to find and document the root cause of any errors from the control group within twenty-four hours of the detection. Without this timely action, it is sometimes impossible to find out what process control collapsed or was missing to cause the inaccuracy. This is precisely the problem with a periodic physical inventory count. Many transactions for the item may need to be reviewed just to attempt to understand the root cause of the error. The objective of a control group is to find the variation or problems in the transaction processing system and fix them. If possible, use the people performing inventory transactions in the control group activity to eliminate the accuracy errors. This will ensure user understanding of any changes to the process, understanding of the process controls, and will ultimately result in process predictability. It means getting process controls in place.

It should be noted here that the recommended thirty to fifty locations is a rule of thumb. As causes are determined, the number of items in the control group should be adjusted to the "right" amount of material or items for the group. As suggested earlier, the "right" amount could be defined as enough locations counted daily to find problems, but not so many as to be unable to reconcile the problems daily. Between thirty and fifty item locations is usually the right number of high-activity locations to check daily, but is not exact as each situation is different. You will have to experiment a bit in the beginning to determine the correct amount. Errors are gems and should be cherished at this point in this process.

Each day, the objective is to determine and document the root cause for each inaccuracy in the control group count. It is important to find the source or root cause of the problem creating the inaccuracy so that action can be taken to eliminate these barriers to high-performance inventory record accuracy. Examples could include:

1. Parts have been moved without the proper corresponding inventory transaction
2. Parts were not received correctly from a supplier
3. The supplier sent a different count than was documented on the packing slip and no verification count was made
4. Lack of proper training resulted in inaccurate picking by a new employee
5. Scrap was made, but not accounted for properly with a corresponding transaction

Once the reasons for errors are understood, Class A criteria require that a measurement document be created to depict and communicate the facts surrounding inaccuracy of inventory records. A Pareto chart works well to display errors and frequency of occurrence visually (Figure 9.1). Pareto analysis allows the emphasis to be targeted at the biggest offenders of accuracy.

(Sample) Reason Definitions:
1. Parts moved without the corresponding transaction
2. Inaccurate picking from warehouse
3. Parts not received correctly from a supplier

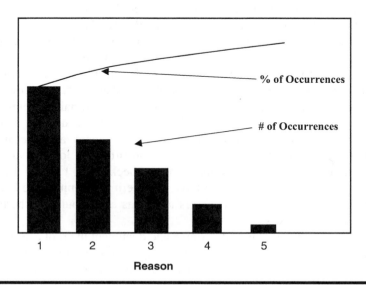

Figure 9.1.   Pareto Chart.

4. Scrap not accounted for in production
5. Production personnel sending parts back into the warehouse with inaccurate transactions

Reasons will vary from company to company, plant to plant, department to department. Your reason codes may be different than those in the sample model. Accurately documenting the errors causing variation in the process is an extremely useful exercise.

Step 3: Determining Root Cause

Once facts are gathered concerning the reasons for record inaccuracy, find the root cause for the reason. Warehouse or inventory managers often find, at this stage, that people do not understand the procedures, process disciplines are not being enforced, suppliers are not holding quantity to the same level of accuracy that they do dimensional tolerances, or there are inaccuracies in the bill of material records caused by a lack of proper standards. All of these can be dealt with easily. The prerequisite is acknowledgment of this variation.

There is no real fail-safe or mistake-proof method for determining root cause. Root cause can be deceptive. The best way to be sure you have arrived at root cause is to determine that action can be driven as a result of knowing the data. Many times, problem solvers will believe they have achieved root cause only to find out that there are more underlying causes beneath the data.

One approach that helps to ensure success in finding real root cause is to apply the proven "5-why" method, also sometimes called "5-why diagramming." This detailed problem-solving methodology is also one of the most simple to use. When reasons for inaccuracy are determined initially, ask "why" five times or until there is clear requirement for corrective action. For each reason, a "reason tree" begins to develop. These reason trees are the basis for understanding real root cause. This exercise can take a lot of wall space or flip chart pages if done correctly. A 5-why diagram can be very easy to use and is very effective in peeling back layers of a given problem or opportunity.

Understanding the Hierarchy of the 5-Why Diagram

The following will give you an idea of the kinds of causes of variation that will be discussed as the 5-why exercise develops.

Sample problem statement: "Inventory is inaccurate."
First-level WHY
1. People have been replaced over the years and training has not been maintained properly

2. Some transactions are not covered by the original intent of the system design, and/or
3. There is not a clearly defined, documented procedure established and published for the correct actions surrounding transactions

Second-level WHY
1. Management has not required documentation
2. There is no central repository for procedural documentation
3. There has been no previous emphasis on inventory accuracy prior to this focus

This "WHY" focus continues until actions can be taken to eliminate some of the sources of variation.

Step 4: Eliminating the Reasons for Inaccuracies

Knowing the reason for inaccuracy leads to the next logical and necessary step: eliminating the reason for inaccuracy. Corrective actions must result from the learning if the control group is to be effective. Often these actions are in the form of improved documentation of procedures, more frequent audits of procedures being followed, and better training in proper procedures and company policies. Sustainable process requires a robust management system.

> Corrective actions must result from the learning if the control group is to be effective.

In Class A organizations, process owners report performance at a management system event called the weekly performance review process. The weekly performance review process is a formal event where management observes progress reports in a predictable meeting held the same time each week and is regarded as an important part of how the business is managed. More on the Class A ERP weekly performance review will be provided in Chapter 14.

Step 5: Ten Error-Free Days

The expected result of root cause analysis and resulting corrective actions will be the achievement of 100 percent accuracy in the *control group*. Step five is to achieve control group accuracy for ten consecutive error-free workdays. This accomplishment will be a significant progress gate. Ten error-free days is an indication that the process is in control — again, for the control group! Since the control group is a representation of the total population of inventory balances, it is a reasonable assumption that these errors, uncovered and eliminated in the control group, were also being committed throughout the facility involving other part numbers.

It is important to understand the realities. Sometimes there can be a heightened awareness of the original control group items after a few days of counting the same parts over and over. Heightened awareness can lead to artificial process control that will not be sustainable over time or necessarily applied consistently over the entire base of parts or inventory.

Step 6: Initiating a Second Control Group

Having the original control group at 100 percent accuracy is great indeed! It is good, however, to verify that the newly installed process control is also in control outside the original control group. Verify your process control success by sample testing with a second control group. The same process used for the original control group (control group A) is repeated. The only difference is the list of materials and locations. This time, companies often choose a larger quantity of items to count since the errors should be less frequent than in the original control group or maybe even completely eliminated. The same rules apply — if the errors are in line with the ability to keep up with daily reconciliation to root cause, the right quantity of counts is being used. Remember that the idea is to find errors as they happen and to eliminate the source of them, if they are still being committed.

Step 7: Establishing a Cycle Count Process

After the second control group verifies process control, a standard/traditional cycle counting program is appropriate. In a standard cycle counting process, a planned schedule ensures that all parts are adequately counted annually. One popular method, and one supported by the American Production and Inventory Control Society (APICS) training materials, has the count frequency of each part dependent on value and/or volume usage of each individual material or part. This frequency is normally scheduled by counting all A items four times a year, B items twice a year, and C items once a year. At this point in the process, the focus again needs to be on the elimination of sources of error.

Many high-performance organizations hold educational sessions on the use of problem-solving tools prior to establishing process improvement teams. This can make the teams much more effective and can lead to faster improvement. Teams that conquer the use of these tools are typically the teams that go on to solve other more challenging problems, allowing the achievement of Class A ERP performance levels not only in areas of inventory data accuracy but also

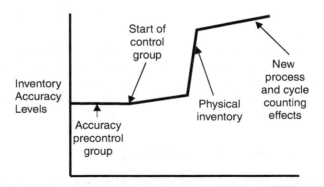

Figure 9.2. The Last Physical Inventory.

in other important process control areas. By using this approach specifically in inventory accuracy and with a control group methodology, the best results can be expected. Employees will be more engaged and will own the results more readily.

PHYSICAL INVENTORY

For many companies beginning their journey to inventory record accuracy, a complete wall-to-wall physical inventory will be appropriate once accurate control groups show that process controls are in place. This is especially appropriate for companies starting out with very low levels of data integrity (pre-control group). Although it is virtually impossible to count a wall-to-wall inventory with 100 percent accuracy, it can be a significant improvement as a starting point for the cycle counting process to begin. The good news is that this is the last physical inventory the company will have to take (Figure 9.2)! Class A businesses do not take annual physical inventories.

Class A businesses do not take annual physical inventories.

ELIMINATING THE PHYSICAL INVENTORY FOREVER

It is the control group (determination of root cause and eliminating each root cause) that allows the effective use of a one-last-time physical inventory. The control group methodology is arguably the most important phase in the 120-day process of achieving inventory record accuracy. It is in these first few days that change begins, allowing the new habits of predictable, repeatable, high performance to be established. Without the control group, it is very difficult to really understand where the variation is coming from and even more difficult to determine what actions should be taken. Strict use of the control group with proper resulting process improvement should forever greatly reduce inventory surprises and/or the associated high costs of expediting and arranging for and funding priority freight costs.

Once robust process is in place to ensure accuracy, the physical inventory normally done annually can and needs to be eliminated. The annual physical inventory is not a value-add exercise. Like inspecting quality into a process, which has long been known as the least-effective method of ensuring quality, the physical inventory is really designed to fix balance accuracy issues once a year. The biggest problem is that it does not eliminate the causes. Also, since it is impossible to count perfectly, especially in a massive factory or warehouse-wide effort, it also can cause some problems that did not exist prior to the count.

Some organizations in the United States believe that the elimination of the physical inventory count once a year is not possible due to imaginary laws or regulations. It is a fact that almost all *high-performance* organizations in the United States legally (and ethically) do not execute an annual physical wall-to-wall inventory. There are some countries that do have regulations governing the requirement, and in that case there is no choice. In companies where regulation requires the count, there is not usually a regulation that requires the updating of the perpetual records postcount. If you are maintaining accuracy, it may be a good idea not to update perpetual records for reasons stated in the last sentence of the paragraph above.

These lessons and others assure organizations that good process pays back in real monetary value and is worth the effort and time spent to establish disciplined practices and procedures. The savings come from many sources and can be very large.

PROCESS OWNER OF INVENTORY ACCURACY

In a Class A organization, the process owner for inventory accuracy is normally the warehouse manager with the most inventory to watch and care for. There are often several inventory owners in various parts of the factory, and each of them can and should serve on the accuracy team, but the leadership position for this important effort is necessary and belongs to the keeper with the most to gain. The decisions made in the redesign of process affect all areas, and one team (including representation from all inventory storage areas) with one leader allows the best opportunity for the right answers to be implemented.

INVENTORY ACCURACY METRICS

Many companies today are surprisingly still at no better than 50 percent location balance accuracy. When organizations address the situation, they find it is not people that have to be addressed first, but almost always the process. People always do what the norm of expectation is.

If the limit on a freeway is a certain speed, but it is not closely monitored or is enforced at speeds considerably higher, and there is not a present and obvious danger, most people will naturally do the maximum the process will allow. This is exactly what happens in the factory and warehouse. People do not see the dangers and costs of taking shortcuts. Many times, they actually think they are helping the cause by reacting to a customer request and shortcutting the paperwork or transaction process. Management even sometimes rewards this "hero" behavior. This is quite contrary to the Class A mentality. Heroes are people who follow the process and who participate in improving it.

The calculation for determining the location balance record accuracy for Class A performance requirements is simple. Normally, location balance counts are done as the cycle counts are administered. Additionally, counts are periodically done in locations to confirm accuracy when suspicions arise. These and any other checks should also be included in the base for calculating accuracy. I refer to these counts as the "foul smell" counts. We count them because they are emitting a suspicious characteristic. Usually, the planning department is the one that gets the suspicious whiff first and requests the counts.

> Usually, the planning department is the one that gets the suspicious whiff first and requests the counts.

The calculation for inventory record accuracy is the total number of accurate location balances (allowing acceptable tolerances) divided by the number of location balances checked within that period:

$$\frac{\text{Total location balances accurate within tolerance}}{\text{Total location balances checked}} \times 100 = \text{Percent performance}$$

There are no tolerances above 5 percent allowable in a Class A environment. The leeway from 3 to 5 percent is to allow inexpensive items, such as pounds of inexpensive resin in a silo or washers worth less than one cent per piece, to have a reasonable tolerance. The good news is that these more open tolerances are rarely required in today's high-performance organizations. The definition of accuracy needs little allowed tolerance if the product can be counted discretely.

If it can be counted, it can be accurate. Inventory weighing scales today are accurate enough that accuracy can be expected even for small parts.

Each controlled area within the facility should be measured separately and results posted visually for all employees to see. At the end of the week, these separate measures should be combined to calculate the overall facility inventory accuracy. This is done by increasing the total base in the metric calculation to include all locations counted. The numerator in the calculation will be the total locations that were counted and found to be correct throughout the facility. This allows everyone in the business to know globally how effective process control is in terms of inventory location balances and yet allows for individual understanding of each area's contribution prior to the consolidated performance view.

BILL OF MATERIAL AND ROUTING RECORD ACCURACY

In addition to inventory location balance accuracy, it is essential to have accuracy in all records used to plan inventory flow and availability. Two such examples are bills of material (BOMs) and routing records. Although it is not the intention of this book to cover all aspects of data integrity requirements, it is appropriate to include a short discussion on these because they have such a big impact on planning material availability and they are included in the Class A ERP criteria. BOM accuracy factors can also impact inventory accuracy.

Components of Bills of Material Accuracy

BOMs are the recipes for the manufacturing process. They also are the records that drive proper assemblage of components for kits in a distribution environment. There are several components of accuracy in BOM records:

- Part number accuracy of parent level
- Part number accuracy of all components
- Unit of measure accuracy and consistency
- Quantity per part accuracy
- BOM structure as it compares to the actual process

The secret to a good accuracy process as it relates to BOM records is easy to understand. There are three applications of BOM records, and they all need to be in sync:

1. What the computer record shows as the BOM
2. What the drawing or specification calls out as the BOM structure
3. What actually is done on the factory or warehouse floor

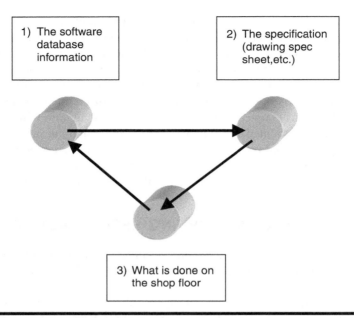

Figure 9.3. BOM Accuracy Match Requirements.

When all three of these files are matched, the result is an accurate BOM record that can be used to manage revisions in the product, flow inventory, and maintain predictable quality (Figure 9.3). BOM accuracy can be especially important when backflush transactions are used to impact inventory balance changes.

The Process Owner for Bill of Material Accuracy

In most Class A ERP organizations, engineering (and therefore the engineering manager) is responsible for the design of products and for detailing specifications required by those products. The BOM is the production translation of this specification into manufacturing terms. This record is used to order and flow materials into place and to assemble the components or raw materials into usable/salable products.

Some smaller organizations or companies that do not focus on new product introduction as a main core process do not have a separate R&D engineering department. In some cases, they might utilize manufacturing engineers or even lead production people to make some of the traditional engineering decisions. In these organizations, someone other than from engineering must be designated as the BOM accuracy process owner. If an engineering department exists in your organization, do not let the people in it talk their way out of this process ownership. It is time for them to join the synergy of Class A ERP.

One of the most frequently heard responses from engineering on this topic is: "I (engineer) can be held responsible for the drawing and maybe even the computer record, but I can't control what the factory floor or warehouse clerk does." That may be true, but in reality, when specifications are not followed, engineering has little chance of succeeding. It is in the best interests of the engineering department to have its specifications followed. As engineering people get involved in the audit process, they often find variation they have not been entirely aware of. This can even result in *higher* quality as engineering designs more manufacturability into the product. The bottom line? Engineering should be the process owner and audit and report performance at the weekly performance review.

When production deviates from specification, it should be noted, with actions resulting to eliminate this condition. When the computer is out of sync with the specification, again, accuracy performance should be reported, with resulting actions. While engineering is not always the one to do the fix, it can be the overseer. Everybody gains and most learn from the process focus.

Measuring Bill of Material Accuracy

The threshold of acceptability for Class A performance in BOM accuracy is 98 percent accuracy. This means that 98 percent of the BOM records must be perfect.

It is not unusual to have several thousand BOM records on file within a manufacturing database. In many organizations, a great majority of the product produced only uses about 20 percent of the part numbers. This can mean that a random sample in work in process would skew toward frequently used records simply because they are in the system more often. The majority of the high-usage parts will be verified quickly and they have the potential to be selected randomly time and time again because of the frequency of use. This is obviously not a valuable use of resource once the accuracy is verified. The following are a few Class A ERP techniques that can be used, some of which deal with the issue of frequently used BOM records and the effect on the metric.

The Simple Random Sample Method

In a simple random sample method, a sample of released BOMs is taken daily until both the confidence and verified accuracy are high. Usually this is done by production workers under the direction of the process owner for BOM accuracy (usually engineering). A random selection of assemblies might be chosen in the morning, and these orders (work orders, shop orders, pick tickets, etc.) are flagged for audit. Workers are trained to scrutinize these jobs for accuracy in all vital areas. A flag on the item master is logged for each BOM as it is checked for accuracy. This eliminates the chance of checking the same BOM too often.

Any error within the BOM makes that BOM inaccurate for the purpose of the measurement. If ten BOMs are checked and two have errors somewhere in the structure, the performance is reported as 80 percent. Since the metric is accumulative over a month, the metric reflects accuracy once the base is built up. The base should be zeroed each month.

Deducting for Accuracy Method

In many environments, after checking BOMs for a few weeks, the accuracy is high enough that errors often are not detected through the daily random sampling. This may be a sign of verified BOM accuracy for released BOMs. At this point, it may very well be appropriate to move to another method of measurement. One way is the accuracy deduction method. Although it might give some statisticians heartburn, it does its job.

When the normal audit method reaches the point where errors are seldom found, the performance is (appropriately) continuously reported at 100 percent. People start to lose any enthusiasm about finding errors. In these same environments, occasionally there is still a "feeling" that 100 percent is not the right number because there are often problems when new products are launched there are often errors when nonfrequently made items are manufactured. This may be a good time to find and post errors as they are detected and stop the process of sampling BOMs on a daily basis. The question of how to measure this is still open.

In this method, each error that is found is given a value of 2 percent and is subtracted from 100 percent. If there is one error, the performance for the day is 98 percent. If there are five errors, the performance is 90 percent and so on. A person with statistical tendencies might be wringing his or her hands nervously after reading that measurement definition. Remember that posting a number is not the objective; the objective is high performance. When errors occur, it is better to have solid recognition of those errors. That forces reporting at the weekly performance review and root cause elimination.

Simple Make-to-Stock Environments

In the more simple make-to-stock environments, there *can* be fewer end-item configurations. In these environments (with less than a few thousand BOM structures), it might make sense to sample the entire BOM database. Often in these environments, there are very few obsolete BOMs maintained anyway, so it is in the best interest of high performance to cycle through the entire BOM record file. This can be done by item number sequence or it may make more sense to audit BOMs in the order of volume usage.

Some companies have hundreds of thousands of BOM records. These companies often utilize a make-to-order (MTO) or engineer-to-order manufacturing methodology that drives the number of possibilities up tremendously. If you are in one of these environments, you may be asking how these companies surface or segregate the right database for audit. A suggestion is to create a firewall between the general database and the ones that are known to be probable manufacturing repeats.

Creating the Firewall

Many organizations have found that by creating a firewall between some of their data, they can greatly reduce the maintenance spending on data upkeep.

> Many organizations have found that by creating a firewall between some of their data, they can greatly reduce the maintenance spending on data upkeep.

The firewall is usually created by defining a "readiness code" field in the item master. Let's say that "P" is the readiness code designated for "production-ready" records. For an item to meet readiness, it has to meet minimum criteria for accuracy.

A leading manufacturer of capital equipment in the 1980s had over 150,000 BOM records. Some of these BOMs were not used frequently, but could be called up occasionally in "same as except for" applications. Businesses in MTO environments often find themselves in this situation. It makes little sense to audit these seldom-used (if ever) BOMs or, in fact, to do any maintenance to these data. This company knew that to ignore these BOMs, without a supporting system for integrity of BOM structures, could create a disastrous result. What this company did was to create a firewall to segregate its data, known from unknown.

In that environment, for an item to get a "P" designation in the readiness code field on the item master, it had to meet the following rigid criteria as a minimum:

1. The record must have a complete and unique part number (not assigned to another item).
2. The record must have a cost standard loaded (even if it is estimated).
3. If it was a parent number (higher level in a BOM), a recorded audit reflecting the latest revisions and part number changes had to be present and available.

4. If it was a manufactured item, an audited routing record reflecting current machine and work centers needed to be loaded. It should also reflect accurate lead times from the routing.
5. The record must include a stocking code indicating if it is a stocked item or an MTO component part.
6. The record should include a lead time if it is a purchased part.

By forcing these requirements for an item prior to allowing it to be released into production, all seldom-used items will have to be converted to a "P" status prior to release. In this example, a "P" also was an indicator that the item would have to be kept up to the latest revision for eternity each time an engineering change was authorized that in some way connected to this item number (unless the production-ready status was changed). This was a big benefit to this company because it was not acceptable in terms of maintenance costs for seldom-used parts or assemblies to be maintained constantly. The company's environment was constant, high-volume change activity.

Some organizations use additional codes in the same item master field for seldom-used items. In this example, let's define the production-ready code on the item master as "X". This will designate parts used "one time only" (seldom). An "X" item, like the "P" item, would have to meet all the criteria for release, but as soon as it is completed, it would again be treated as a nonmaintained item. Engineering change would not update this item until it was called into action again, if ever. In this example, the firewall procedure would stop any item *not* designated as a "P" from being released or ordered. Even an "X" item would have to be reviewed every time it was released. This saves on engineering maintenance and costs when items are often "one time only." Anything with a "P" status has to be maintained always and kept current through every engineering change connected to it. A "non-P" item is behind the firewall and has to be reviewed each time (if) it is used in production.

This process allows the engineering department to invent and utilize the database any way it wants without any risks to production or inventory control. As long as engineering stays behind the firewall with its experimental BOM designs, it is not bothering anyone. Many organizations find this a rewarding system in more ways than one.

Routing Records

Routing records are the process steps or maps that materials follow through the process of being converted from raw material into finished goods. In some organizations, routing records can be very important, and in other environments, they are not important at all other than for cost standards.

In regulated organizations such as those in the pharmaceutical industry, the routing is important as a definition of process. It has to be accurate and has to be followed to the letter due to government regulation and liability concerns. In these companies, a lot of attention is paid to routing records.

In traditional metal fabricating companies and long cycle time processes, routing accuracy is also important. Routing accuracy in these businesses must also be in the 99+ percent range. In these environments, raw material goes through several conversion operations and may travel reasonably long distances through the process. Here, routing records are necessary for the planning process. These routings might read something like:

Operation	Instruction	Machine Center
010	Saw to length	7230
020	Flamecut notch	7620
030	Deburr/grind	3410
040	Form	4620
999	Receive into stock	9999

These routings tend to support longer cycle times and more queue time as jobs move through functional departments such as "saw department" or "lathe area."

Today's high-performance organizations that utilize lean methodology are opting for short cycle times where material travels only short distances and operations are close in proximity and batches are small. In these environments, the same routings might read something like:

Operation	Instruction	Cell
010	Make part	3
999	Move to cell 7	7

In these environments, no labor transactions are done because the time to do the process is extremely short. In these companies, only inventory transactions are recorded.

Reviewing the record accuracy of routings is done much like that of BOMs. Engineering, usually manufacturing engineering or process engineering, is the responsible department and process owner for the accuracy audit and reporting. In smaller businesses where people wear many hats, again, someone from production would probably be the right person for this process ownership.

Many of the newer computer systems have combined the BOM record and the routing record into one integrated record file in the system. These are usually referred to as bills of resource. In this application, the BOM structure is expanded to include the process in which the component part is used. Sequencing and timing can be better planned. Again, this is not as beneficial in process flow or lean environments where inventory moves fast through several processes.

SUMMARY

There are other elements of data that drive decisions daily in manufacturing organizations. Each field in the item master could be a topic of discussion in this chapter. The most important ones have been covered. As data accuracy is identified as a root cause for other metric performance issues (and it often is), actions must be driven to eliminate the causes for the process variation. Process ownership for each field in the item master should be assigned at the start of the Class A implementation, even before the causes of problems are identified.

10

CLASS A ERP PROCUREMENT PROCESS

Supply chain management has received a lot of attention in books and seminars in the last few years — with good reason. Suppliers are the lifeline of most businesses. Whenever their performance is anything but synchronized with the procurement schedule, unnecessary costs are incurred. Many poorly managed businesses run their supply chain management through expediting and priority freight. Class A businesses do not work that way. Good supply chain management means that the suppliers of greatest importance share information daily, sometimes hourly. Whenever changes are made to the schedule, suppliers are involved in the decisions, particularly when lead times or rules of engagement are violated. Figure 10.1 was also used in Chapter 7 on master scheduling. This illustration will become more and more meaningful as the discussion on the supply chain continues.

In a Class A ERP organization, there should be little schedule movement within the fixed fence. Any changes to the schedule within this short time frame need to be cleared with the supplier. Remember that this time frame does not have to be very long. The statistics in Table 10.1 suggest average time frames for the fixed fence in an unscientific survey of high-performance clients I have worked with.

It becomes obvious that most businesses are tightening their fixed fences to quite short periods, at least compared to just a few years ago when "frozen schedules" were often quoted (but seldom exercised) at two-week levels. This

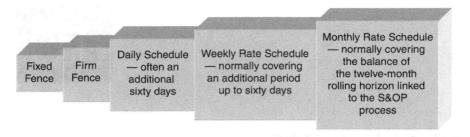

Figure 10.1. Time Fence Norms and Flexibility Built into the Planning Horizon.

means that suppliers are often forced by market conditions to determine inventory strategies to meet these requirements. Inventory strategy is one of the first decisions in Class A businesses. After all, all businesses are suppliers to someone, and Class A is a process for all businesses, especially manufacturing firms.

If 60 percent of manufacturing firms have schedules with their customers that are only fixed for about forty-eight hours, some thinking and planning has to go into being ready for changes in the forty-ninth hour. Not many businesses have the luxury of full manufacturing cumulative lead times at less than forty-eight hours. That requires inventory strategies that buffer inventory somewhere in the supply stream. If we review Figure 10.1, it becomes obvious that without good forecasts and handshakes that establish agreed-upon variation tolerances within the various time fences, things could get pretty ugly. Ugly equals cost, and in many businesses, that is exactly the case.

In Figure 10.2, flexibility is shown as an important need for a handshake agreement at the beginning of the relationship. In this example, for a particular product family, the supplier knows exactly what the requirements are for flexibility of schedule. This eliminates many of the possible surprises. Table 10.2 shows the detailed agreed-to rules of engagement. If the supplier in this example

Table 10.1. Fixed Fence Time Frames

Time Frame	Percentage	Cumulative Percentage
One day	35 percent	
Two days	25 percent	60 percent within two days
Three days	5 percent	65 percent within three days
Five days	20 percent	85 percent within five days
Two weeks	10 percent	95 percent within two weeks
Over two weeks	5 percent	

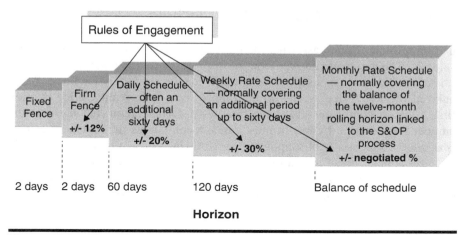

Figure 10.2. Time Fence Norms and Flexibility Built into the Planning Horizon.

Table 10.2.   Flexibility Requirements Table

Time Frame	Schedule Fence	Lower Limit	Upper Limit
0–2 days	Fixed fence	100 per day	100 per day
3–4 days	Firm fence	88 per day	112 per day
5–64 days	Flex fence	80 per day	120 per day
65–125 days	Forecast fence	70 per day	130 per day
126–365 days	Open fence	No limit	No limit

has a cumulative lead time of twenty-four days, inventory strategy comes into play immediately.

WHY DEVELOP PARTNERSHIPS WITH SUPPLIERS?

In recent years, lean thinking has taught us that costs eliminated anywhere in the supply chain are good for everybody in that chain. Partnerships are special working relationships that allow and foster sharing of technology, forecasts, and financial information. "Partnership" in this context does not refer to the legal term, nor does it suggest any special legal implications, although they are not excluded from the process. Instead, we are talking about trust and shared goals. When members within the supply chain trust each other and work together closely, there is great opportunity for elimination of waste and duplication.

SELECTING SUPPLIERS FOR PARTNERSHIPS

During training, I often ask what the most important measurement for supplier value is. The automatic answer from most professionals is normally "quality." It is a conditioned response; everyone has evidently read the same book. When quizzed further, "price" seems to be a close second. In reality, price is often first. Class A companies also have price as a consideration, but it is not the number one criterion by far. In most cases, a continuous improvement culture is as good a sign as any of supplier value. Knowing that a supplier is constantly after yet another notch in quality, responsiveness, or cost reduction gives most customers a feeling of confidence. Probably the real number one factor, however, is trust or, said a different way, reliability.

Criticality

When considering partnerships with suppliers, there are a few criteria that need to be considered, not the least of which is criticality. Some suppliers are easy picks because they are joined at the hip with their customers in terms of shared technology, financial information, and forecasts. In some cases, it is not that easy. In some commodity markets, the best suppliers supply both you *and* the competition. This is a definite consideration. Certainly, criticality of the supplier is a number one qualifier. Single-source suppliers, which by definition are extremely critical, often should be on the list of first possibilities to be evaluated.

Reliability

Before entering into a partnership agreement with any supplier, reliability needs to be both examined and experienced. Being able to turn attention away from a supplier and have confidence that the supplier will deliver when and what you need is a huge asset and a load removed from your daily obligations. Reliability comes in many forms. The Class A ERP criteria look for:

- **Quality** — Reliability of both quality process and product is simply table-stakes today. Without it, a supplier should not even be on an approved list, to say nothing about a partnership consideration. Quality of process is defined by having repeatable processes, management systems to ensure compliance, and metrics in place to warn of process variation before it impacts specification compliance. Class A organizations measure first-time quality. First-time quality is a good indicator of quality process and should be part of any partnership consideration.

■ **Integrity of promises** — Reliability of process is only meaningful if it includes reliability of promises. Promises in this category would include dates met on new product design commitments for components, quality improvements, cost reductions, and not least important, delivery of product to the customer's door. Along with this comes the trust that when problems arise (and they do, even in Class A organizations), there will be an early-warning system to allow plan realignment with minimal loss.

■ **Responsiveness to schedule changes** — The supplier is not the only potential reason for schedule changes. More often, customers cause the process variation. If you think about it, the list of requirements discussed in this section applies not only to our suppliers, but to you (as a supplier to your customers) as well. Everybody is a supplier. It is *your* responsiveness to customer need that requires equal or better responsiveness from *your* suppliers. Is this getting confusing? The truth is, the farther you look down the food chain, the more responsive the parties need to be.

Competence

Similar to reliability, competence has many faces. There is no reason to develop partnerships with suppliers that lack trustworthy competence. While being reliable is good, being reliably competent is great. There are, however, a few expectations in this space:

■ **Technical expertise** — If a company makes the highest-quality goods, it should also understand the technology behind those products. Only a few years ago, most businesses could afford full R&D staffs. With today's competitive markets and lean thinking, most companies are relying (at least to some degree) on suppliers for technical assistance in the suppliers' fields of expertise. This only makes sense. Total cost is incurred throughout the supply chain, not just in your factory or business. If you also have engineers designing a component, there can be a duplication of resource within the supply chain. The easiest component-design-related resource and associated cost to eliminate is yours.

■ **Financial stability** — Supplier profitability has not always been on the top of customer priority lists. It is not generally the focus of everyday concern. Companies that have experienced the bankruptcy of a single-source critical-component supplier think differently. Financial stability is the enabler that allows design improvements, capacity investments, and risk taking for the sake of improvement.

- **Improvement history and improvement plans** — With competence comes the willingness and tenacity to stay at the head of the pack. All great companies have a continuous improvement track and measure progress and results regularly. Some progress is visible to the customer, but not all of it. When considering the possibility of a partnership handshake with a supplier, ask for evidence of its internal process improvements.

Location

This topic is controversial at best. I have had several U.S. clients that recently decided to source offshore. The North American Free Trade Agreement (NAFTA) was part of the reason, but China's favored nation status with the United States was also a driver. Some people would probably not include this topic on the list of qualifiers for determining potential partnerships. There are no firm and fast rules about the importance of supplier location. There are always trade-offs. Distance equals cost, and adding either is bad.

Location is on this list not because there are recommendations that always apply, but because the risks are many and the rewards are not always as significant as planned. As human beings, we know some processes well and yet others not as well. We are good at getting components across great distances (thanks to rail and ships) at relatively low cost. We are also good (thanks to priority freight companies) at getting components across great distances quickly. But we are not good at getting components across great distances quickly *and* cheaply. The "cheaply" part of freight transportation is not always baked into the planning cake initially. I can reference several situations where companies have sourced offshore only to see their worst nightmares happen. I am working with one such case right now. In this case, a component family was moved from Mexico to China, and unplanned problems began immediately. I am not a fan of cross-world components generally, but often end up with that solution in specific situations. The bottom line is that too often people do not do enough homework and become entranced by the taste of (seemingly) getting something for nothing.

> I am not a fan of cross-world components generally, but often end up with that solution in specific situations.

There is another way to look at location as a competitive advantage. One client I work with is a blow-molding company engaged in lower-volume packaging needs, mostly for the food and drug industry. Opposite from the offshore approach, its strategy is to build plants across the street from its customers. In

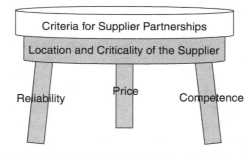

Figure 10.3. Three-Legged Partnership Criteria Stool.

the past few years, it has built two new plants and is getting ready to break ground again. This strategy has created and funded the company's growth.

Many organizations that are moving offshore for components are starting to require buffer within a radius of the customer's facility. This, of course, comes at a price also. Again, there are always trade-offs!

Price

Notice that price is last on the list. This does not mean it is not important; after all, it *is* on the list. While price is a very important element, it is just not as important as reliability and competence. The three-legged stool (purposely leaving location out of it) would be pretty wobbly without any one of the legs (Figure 10.3). Price does not need to be the lowest in the marketplace. Price needs to provide good value. Very few high-performance companies have become successful on the basis of price only. Most of the best companies in terms of growth and stakeholder return are not the lowest priced in their markets. The common denominator is value. Value is built from product quality, service behind the product, and people behind the products and services.

IMPLEMENTING SUPPLIER PARTNERSHIPS

Once suppliers are chosen for partnership development, the next question that needs to be answered is how to go about successful implementation. Project management is a proven process, and each supplier partnership implementation should be a separate project with a team leading, measuring, and supporting the progress.

■ **Define the scope of the project** — As with any project, the first step is to define the scope. When does the group get to buy the pizza and celebrate victory? How long does the team have to get the job done? What are the deliverables from this partnership in terms of information sharing? What are the rules of engagement once the partnership is in place? What happens in unplanned situations?

■ **Choose the team** — The team makeup is very important. It cannot be made up exclusively of lower-level individuals, although there is good reason to have people from within each company who understand the details, including how the information systems work.

■ **Typical partnership implementation team** —
  □ Sponsor — Customer VP of operations (or VP of procurement if there is one)
  □ Team leader — Customer purchasing manager
  □ Team members — Customer materials manager, customer planner for this product family, supplier account manager or salesperson, supplier master scheduler, supplier product manager for this product family
  □ Support personnel — Customer information technology manager, supplier information technology manager

■ **Develop the implementation plan** — The implementation plan is no different than for any project. The elements of the plan should include:
  □ Collect data that describe the current "as is" condition
  □ Define the expectations in measurable terms — the "should be" state
  □ Develop plans to close the gap
  □ Process map the information flow, product flow, and technology or knowledge flow that happens both today and in the future model

■ **Create and assign actions in Gantt chart format** — Document all the expectations. Do not lose original promise dates even when rescheduled. Maintain two columns: original and revised dates.
  □ Drive actions to move the relationship to the desired state
  □ Measure the results and adjust as necessary
  □ Document the policies and rules of engagement
  □ Monitor results

■ **Insist that the gates are met prior to moving to the next step** — Too many teams, under pressure to enjoy projected savings, will shortcut steps or even skip them. In my experience, it is better to be a little late than a lot wrong. Shifting the supply chain is a high-risk move that needs to have the best project managers and should not be delegated too far down in the organization.

- **Celebrate** — Too many teams forget to celebrate success. This does not require fireworks, but if the desired outcomes were met, some recognition is in order. The celebration spotlight helps educate others in the organization on what good behavior looks like. It is not just about the team members, but also the rest of the organization. There will be a lot of additional partnership agreements to implement.

ALIGNING THE SALES AND OPERATIONS PLANNING PROCESSES

While the partnership develops, one of the most rewarding efforts in terms of information flow is the alignment of the sales and operations planning (S&OP) processes from one business to the other. Figure 10.4 illustrates the alignment using the ERP business model as the schematic.

The information streams coming from the S&OP processes are the most up-to-date information available and also the more reliable and accurate. The information is never perfect because it is only a prediction (forecast), but it is the best available. It should be shared. Depending on the size of the product family and impact of the supplier, in some organizations the suppliers actually attend the S&OP meeting for their particular product family. Much more com-

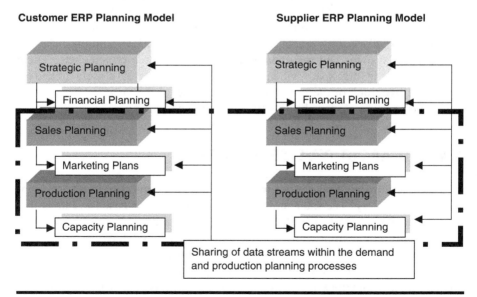

Figure 10.4. Aligning the ERP models and S&OP with Suppliers.

monly, the updated information is forwarded as soon as the meeting is adjourned and plans blessed.

INFORMATION SHARING

There are many types of data that businesses within a food chain can share profitably. The possibilities are endless, but here are a few ideas:

- Demand plans
- Production plans
- Pricing information on shared commodities (negotiate together on things like steel or resin)
- New product introduction schedules and expectations
- Quality control techniques
- Changeover improvement techniques and expertise
- Technical information on products manufactured
- Competitive analysis
- Strategic goals as they relate to shared interests
- Business imperatives as they relate to shared interests

Depending on the specifics, the information sharing can go well beyond this list of obvious topics. Sometimes companies will share other nonproduct-related information, just to help a supplier be as professional as possible. This might include things like safety practices, Six Sigma training, lean training, Class A management systems and methodologies, human resource policies, etc. There are few limits in this space when it makes sense to be as closely linked as possible. If we think of the supplier as an extension of the business, more ideas can come to mind as possibilities to increase synergy.

A WORD ABOUT REVERSE AUCTIONS

Reverse auctions have taken root in most markets at some level and are probably around to stay. The interesting thing about these web-based events is that they seem to go against everything I have learned in my thirty years of developing supply chain improvements. Nonetheless, they do have a place. In a reverse auction, invited attendees share a web gathering for the sole purpose of "bidding" on a customer's business. Bidding consists of lowering the agreed price. The specifications are supplied prior to the event and usually include starting price and quality specifications along with drawings of technical requirements. More recently, companies are beginning to build flexibility and responsiveness requirements into the auction criteria. This is an excellent use of the process.

In an unscientific study conducted within our client base, it seems that more than 80 percent of the auctions result in keeping the same supplier after the auction as before. This has led some suppliers to decide not to participate when reverse auctions are scheduled by their best customers. (From the supplier standpoint, this is a little risky, and yet in the right circumstances, it is the right thing to do.) The reverse auction is an intriguing topic and a bit like poker. The rules of this game have not been completely flushed out. The strategies for reverse auctions will continue to be honed over the next couple of years. Most high-performance companies today are using reverse auctions to some degree in certain applications. It is not always the best thing to do to your best supply chain partners, however.

> Most high-performance companies today are using reverse auctions to some degree in certain applications. It is not always the best thing to do to your best supply chain partners, however.

LOGISTICS AS A COMPETITIVE ADVANTAGE

Logistics is a huge expense within any supply chain. No matter what you produce, parts of it have been moved somewhere, sometime recently, probably by truck and many times by additional modes such as rail or ship. Transportation is a huge opportunity for waste elimination and lean thinking. Any time an empty truck can have something in it instead of deadheading or every time distance can be minimized in transport, improved cost is normally the result.

Today's freight haulers are trying to differentiate themselves just the same as every other business. These organizations can become partners as well in limiting supply chain costs. Sometimes it is the smaller regional freight haulers that have the most flexibility, so do not eliminate them in your testing. If there are clusters of suppliers in certain regions, "milk runs" can help eliminate costs by consolidating several suppliers' shipments into one trailer load. Process mapping is the right tool to use to tackle this normally complex opportunity. Here is a short list of activities to remember as you focus on logistics as a customer service and/or cost-reducing opportunity.

- Understand the total transportation requirement. Look at and map the total freight picture, in and out of your facility. Look for activity clusters where freight could be combined with other shipments.

- Review the inventory strategy of each product family and make sure there are no opportunities there to combine strategies.
- Engineer common shared components wherever possible.
- If possible, do not have inventory made ahead of time and keep all inventory in the supply chain moving. If it does not need to move, maybe it should not have been made. Increase inventory velocity.
- Sometimes consigned inventory is a good solution that saves freight. This has to be weighed against the cost of inventory.
- Optimize geographic stocking locations if using distribution centers. Make sure they are aligned with the customer base and, if appropriate, the supplier base.
- Use of reusable containers is often a cost-saving option.
- Evaluate cross-docking versus warehousing of inventory for cost opportunities.
- Stocking generic end packaging for the specific need can create flexibility in inventory.

In many instances, there are opportunities for the freight hauler to be involved in supplier partnership implementation. Even being made a member of the team may make sense. Many high-performance companies have included these logistics providers in the problem-solving efforts that have resulted in new improved solutions. Remember that they are often an important supplier in the overall cost and service picture. Again, do not make the mistake of going with best price only in this decision!

SUPPLIER PERFORMANCE RATINGS AND CERTIFICATION

Supplier rating and certification may be one of the most rewarding and fun topics in this chapter on procurement and supply chain management. Without a doubt, it has a lot of impact on procurement success. Class A ERP is built on metrics and management systems, and this is the heart and soul of the Class A ERP supply chain process. Metering supplier effectiveness is best done using three elements: quality, delivery, and service. The objective is to keep subjectivity out of the equation as much as possible, although supplier ratings will not be completely nonsubjective.

Rating System for Existing Suppliers

95 percent sustained = Certified supplier
90 percent sustained = Qualified supplier
80 percent sustained = Approved supplier

Rating Category	Weight
Quality	30 percent
Perfect orders	
Packaging quality	
Count accuracy	
Delivery	40 percent
Promises met 100 percent	
Flexibility in schedules	
Value-add services	30 percent
Short lead time	
Technical support	
Consigned inventory	
Other value-add processes	

Application of this model normally includes written criteria for each level within the areas of quality, delivery, and value-add services. This can be developed in conjunction with a supplier or done independently. It is always best to have the supplier review the criteria prior to implementation. If small adjustments are necessary, ownership in the end product will be improved.

There will be times when a supplier is "the only game in town" and approval criteria are not met for full authorization as a supplier. Sometimes because of business need, jettisoning the supplier is not an option. This is not a favorable position to be in, but it is not terribly unusual. The mission objective is to get out of this position as soon as possible. The options are not always apparent, but some that should be considered include coaching an alternative favored supplier to provide the additional product line, looking out of the area for an alternative supplier, adding the capability to your own facility, or coaching the existing supplier to improve. Class A ERP certification can be powerful in this regard. Some clients I have worked with have actually paid for the education of their suppliers to enjoy the benefits of improved performance. There is no magic potion here, but the lesson has been learned by many organizations over the years. Do not accept poor suppliers any longer than absolutely necessary. This is an easy thing to say and not always easy to accomplish, but it must become a business imperative when in this vulnerable position.

Within the supplier rating system, there is normally a maturity profile that rates progress of the supplier base. There is some leeway in these ratings evidenced by many high-performance organizations. Table 10.3 gives some suggested categories and rating criteria for a supplier maturity profile.

> **Sometimes because of business need, jettisoning the supplier is not an option.**

Table 10.3. Maturity Profile Specifics: Supplier Status Characteristics

Certified Supplier	Qualified Supplier	Approved Supplier
Minimum of three perfect receipts	No rejected material for six months	Last receipt perfect
No rejected material for six months	Quality, delivery 95 percent	No rejected material for three months
Quality, delivery 95 percent	Monthly performance review	Quality 95 percent
Value-add services 95 percent	Annual audits	Delivery 90 percent
Audits optional		Audit every six months

INTEGRATION OF LEAN AND SIX SIGMA IN PROCUREMENT

As with most Class A ERP topics of discussion, the integration of lean and Six Sigma is a no-brainer. True Class A ERP cannot be sustained without this thinking.

- **Lean** — The elimination of waste is enjoyed quickly within supply chain management as lead times are reduced. Every time a day is struck from the lead time of a widely used product, the associated inventory for that twenty-four hours is eliminated as well. Using the technical resources of suppliers instead of funding them in-house is another example of the lean principles being utilized. Each time the quality is improved and process variation from suppliers is reduced, lean principles are engaged. Without lean, efficient supply chain management cannot exist.
- **Six Sigma** — Project management has a lot of opportunity within the supply chain management space as well. Improving quality, eliminating inventory, implementing supplier partnerships, and coaching setup reduction at the supplier site are all examples of project possibilities. When black belts (people who have high project management and problem-solving skills) are engaged for the accomplishment of these tasks, the supply chain becomes more effective quickly.

PROCESS OWNERSHIP IN PROCUREMENT

The procurement manager is the natural process owner for the Class A ERP procurement process and for supply chain management. In some larger organizations, there might be a VP of procurement. In that case, he or she would be the Class A process owner for this area. Duties of this process owner include tracking metrics, reporting performance, root cause analysis of process variation, driving change through actions, communicating the performance to the suppliers, following up on audits and supplier assessments, managing supplier partnerships, and tracking practice to the current policy. This is no small task and one that is left to strong-willed skillful leaders in high-performance organizations.

CLASS A METRICS IN PROCUREMENT

The metrics in Class A procurement focus on system linkage and synchronization. Although supplier metrics seem to be a significant part of supply chain management, the metric emphasis in Class A is larger than just looking at the supplier. The real supply-chain-related opportunities in Class A come from making sure that the business system is maintained to the latest master production schedule (MPS) and that the suppliers are absolutely in line with the latest schedules.

> The real supply-chain-related opportunities in Class A come from making sure that the business system is maintained to the latest MPS and that the suppliers are absolutely in line with the latest schedules.

When the master scheduler changes something in the MPS, unless the suppliers are hardwired to the revisions, little gain will be seen from making the change in the MPS. For that reason, the procurement process metric in Class A is the percent of complete orders that are received on the day that they were required.

Class A Procurement Process Metric

The proper Class A procurement process metric is to measure the percent of complete orders or releases from purchase orders that arrive on the day or at the hour (depending on the schedule interval) that the parts were "required." For this metric, the definition of "the day or hour the parts were required" will be found within the ERP business system. The required date/time is the latest requirement date/time in the system at the time the material was received. The definition of "complete" means exact quantity to the latest order or release and 100 percent usable parts — no rejects. Any reject makes the order a miss for the day. If there are multiple stockkeeping units on one purchase order, the full order is the definition of the requirement. This metric is at a higher standard than many procurement process measurements, but the lessons learned from this will be very beneficial to the business. Many times, the supplier is not the problem at all; the schedule stability is. Without this view, it is more difficult for the purchasing professionals to get their point across when suppliers are being "jerked around" by schedule changes. It is, after all, very easy to blame the supplier.

The traditional supplier metric of measuring the integrity of the last promise from the customer is a reasonable lower-level metric and often is a good source of data for the procurement people in manufacturing. Alone, this metric does not meet the correct metric requirements for Class A criteria and does not get into enough detail to consistently point to the root cause of being out of synchronization.

The next chapter will deal with the shop floor and issues encountered there. We will find that the opportunity to use lean and Six Sigma is not only frequent but constant.

11

SHOP FLOOR CONTROL

Organizations that have Class A ERP processes in place find that real cultural changes driven from the Class A standard always start at the top. Good solid management planning along with prioritization of such activities as measurement habits, root cause analysis, and a robust management system are very influential on business behavior. That needs to be part of the thinking as the discussion changes to the shop floor. When top management carries its part of the load, this part of Class A ERP gets much easier and more predictable.

I am reminded of a request I received from a Fortune 50 company a few years ago. The company wanted me to visit a plant because "the shop floor was all screwed up." As soon as I heard the phrase, I smiled to myself because I did not want to reveal what I was really thinking before I had proof. It did not take very long to get that proof. When I entered the plant later that month, I was put in a holding pattern because the plant manager was in an emergency meeting. This was just a hint of what was to come. After about a half-hour wait and a subsequent discussion with the materials manager, I finally met the plant manager. She was an experienced woman with an obvious love of communication — a lot of it! I am not talking about formal communication of plans and actions and schedules. I am talking about short, constant bursts of pseudo "progress" reports. This plant manager had a walkie-talkie, a beeper, *and* a cell phone. I soon found that it was impossible to have a discussion with her for more than a couple of minutes without one of her communication devices ringing, beeping, or screeching.

Her instructions were brief. "I would like you to spend some time on the floor and figure out what the problem is there," she said. I agreed to do so, and for the next week I spent time each day on the floor talking to and observing the activities of supervisors and supporting personnel, *including about forty expeditors.*

The product they were manufacturing was about as large as a full-size Buick. They averaged only a few units (single digits) into shipping on Mondays, but averaged twenty-seven a day regularly on Fridays. As you would imagine, that is not how these units were scheduled. The problem was clearly management — lack of it. There was no discipline in making or carrying out schedules. Past-due shop and purchase orders were commonplace, and the factory motto was "whatever it takes." This was the model of chaos!

> The problem was clearly management — lack of it. There was no discipline in making or carrying out schedules. Past-due shop and purchase orders were commonplace, and the factory motto was "whatever it takes." This was the model of chaos!

In high-performance Class A processes, there is very high discipline in both planning and execution. When management insists on accuracy, root cause is more likely to be found and dealt with. When management insists on effort only (as opposed to focus on process compliance and discipline), that is often the result — a lot of effort only. Effort without vision, focus, and discipline is often wasted energy. I went on to explain to this plant manager that the floor was not the problem. Data were inaccurate, schedules were not well thought out and matched to capacity, schedule stability was nonexistent, and schedule-related communication was not done through the business system, but instead through notes and shouting on the walkie-talkie. There was no effort to get a handshake on rules of engagement or a sales and operations planning (S&OP) process. The plant manager did not want to hear this, and it was the last time I was in that facility.

> The plant manager did not want to hear this, and it was the last time I was in that facility.

That was a few years ago, and the saddest part of this story is that this plant is now closed and the products are produced offshore. Class A ERP process and disciplines could have *at the very least* delayed this move several years if not longer.

Figure 11.1 depicts the entire Class A ERP process. It is the top quarter of the business model (circled) that makes or breaks a business in terms of efficiency and effectiveness. When there is good top-management planning, the shop floor activity is not only effective, but easy!

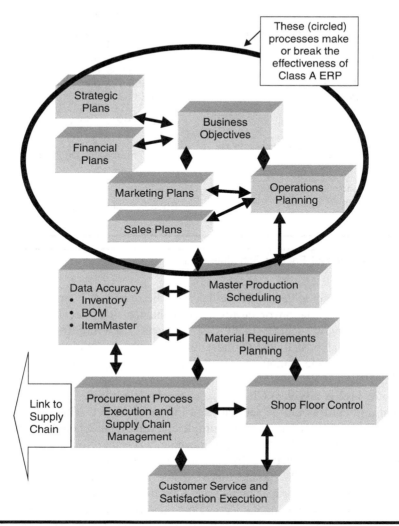

Figure 11.1. Class A ERP Business Model.

COMMUNICATING THE SCHEDULE TO THE FACTORY

The master production schedule (MPS) is the driver of all schedule activity in the factory. In a Class A organization, the MPS and shop floor schedules are identical at the beginning of the week. These schedules are also identical on Wednesday morning, Friday morning, and any other morning you want to

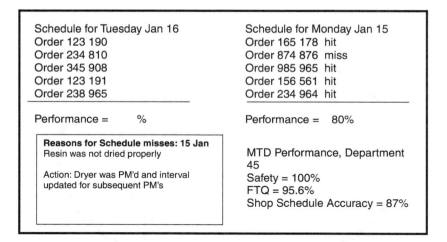

Figure 11.2.   Visual Factory Schedule Board.

discuss. It is no coincidence. Schedules must be linked. It is not because of a lack of process variation or customer misbehavior. It is because the master scheduler insists on accuracy in a Class A process. Each morning, the schedule is agreed to and posted for all the employees to see. This "visual factory" technique was made famous by Toyota and is a simple idea based in lean thinking.

Figure 11.2 is a typical Class A factory schedule board. This type of communication board would be seen at each line or work cell within the plant. It allows every worker to understand completely what their schedule is and what their past performance has been. Each morning, a special effort is made to bring attention to this board as the daily schedule is posted and agreements are made on the day's tactics. By having this schedule as the center of focus and by having little tolerance for inaccurate schedules, discipline becomes the habit and results become very predictable. Every time there is a schedule miss, the serious question has to be asked, "What can we do to stop it from ever happening again?" The question is answered by the process owner for this line. Sometimes it is the master schedule, sometimes it is the machinery reliability, sometimes it is easy to eliminate, and sometimes it takes some horsepower. In any case, it needs to be addressed or it will continue.

Getting shared goals on the floor and a tenacious attitude toward accuracy will get any plant floor on the right track for Class A performance. One special management system event in the factory required by Class A is the daily walk-around. Done on the factory floor, this meeting is management's morning opportunity to get updates from each line. The plant manager, master scheduler, and possibly the manufacturing engineering manager along with other willing or interested parties walk through the plant and stop at each line or work cell. Using the visual factory schedule board as the centerpiece at each stop, the management team will receive an update each morning from each production team. Normally, the team leader gives this update. The format is kept consistent: review yesterday's performance to schedule, review actions completed to eliminate any process variation experienced, and review today's schedule and any anticipated issues.

In many businesses, the first barrier to success is not accepting or acknowledging failure. In these confused organizations, there may be a "can-do" attitude when often in reality they "can't do." Attitude is a great enabler, but it is not the only requirement for success; good process design is also necessary. In such cases, the requirement is good communication and a handshake between the master scheduler and the shop floor, balanced by appropriate rules of engagement. The schedule needs to match reality for the supply chain and process to be in sync. I was recently in a factory where the schedule was consistently above the capacity. The result was twenty truckloads of unneeded material and no place to put it. Waste.

THE ROLE OF MATERIAL REQUIREMENTS PLANNING

Material requirements planning (MRP) is a planning tool. Firm requirements are not the same as planned requirements. Whereas kanban and pull systems are designed and efficient to take care of firm requirements, MRP is a tool that still has its place in managing the planned requirements. At some point, the planned orders become firm, and that is when MRP adds little value and lean steps up to the plate to keep schedule adherence disciplines and keep waste out of the process at the same time. Once the orders are firm and the MPS has authorized the start of work on a requirement, there is no need for MRP (Figure 11.3).

In one of my earliest examples of this, a capital goods manufacturer had customer-order steel requirements every day that had to be welded to make the main infrastructure of the product. Both the height and width of the final product could vary depending on customer-specific application, so there was no reason to build this product early. To do so would mean forecasting requirements and stockpiling an infinite number of possibilities. It was a no-brainer to use lean techniques. This was even before lean had made the circuit as the latest thinking. Going even beyond this, the other smaller components of the process that welded to these larger steel members would normally have been planned by MRP and pulled from stock as needed. The process was instead migrated to lean. Kanban (although it was not called that back then) was used

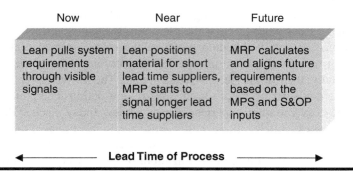

Now	Near	Future
Lean pulls system requirements through visible signals	Lean positions material for short lead time suppliers, MRP starts to signal longer lead time suppliers	MRP calculates and aligns future requirements based on the MPS and S&OP inputs

◄———————— **Lead Time of Process** ————————►

Figure 11.3. MRP Integration with Lean.

to pull the manufacturing of the components from the fabrication area. Quantities were based on firm orders. This worked great with short-lead-time components. Fabrication became flexible and was able to respond quickly to these requirements. Bin size or storage area was determined by the lead time of the component.

The bill of material for this weldment was loaded in the system, but the planning code on the item master indicated to the planner to ignore any signals. This gave full control (beyond the MPS) to the floor for all of its requirements. The larger steel members were ordered daily by factory workers in the process directly from the supplier and were delivered the following day cut to exact length. This eliminated all inventory in the process except for the units actually having work done within twenty-four hours. It also helped to eliminate thousands of dollars of inventory.

The MRP system's part in this planning process was simply to give the raw material suppliers visibility out, twelve months rolling, to the MPS horizon. Because it was driven by a robust top-management planning system, an S&OP process, the information was always the best information the supply chain could get. Nothing from MRP was firm to the suppliers. It was visibility to planned expectations in the future. This is helpful for capacity planning within the supply chain. It might not be perfectly accurate, but it was to the latest thinking from the business management. MRP and lean methodologies have a solid partnership in high-performance businesses.

FUNCTIONAL MANUFACTURING VERSUS PROCESS FLOW

Not very long ago, manufacturing departments were always arranged by function. Welders were located together, lathes were all in the same department,

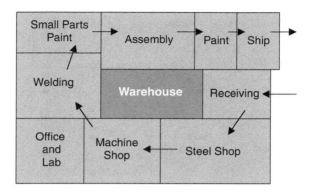

Figure 11.4.   Functional Manufacturing Flow Layout.

milling machines were set side by side, etc. It just seemed to make sense. This was prior to the acute awareness of waste developed from lean thinking. In today's high-performance organizations, there is a much more educated view of material flow.

At The Raymond Corporation in the 1970s, we had what we considered to be good process flow. I remember giving tours through the plant and explaining that material started in the steel shop and progressed to shipping in a logical layout. The plant looked very much like Figure 11.4.

While the factory flow in Figure 11.4 may seem like a logical path for process flow (and during tours it was even described in that way), in reality it was not. It took an exercise of tracking actual material flow through the process to understand the opportunity. On a factory layout drawing, we tracked the flow and marked the path on the paper. What we found was something like Figure 11.5. Material actually flowed around and around the shop. If we dropped a string on the path, it probably would have taken miles. We did not measure it. This was an eye-opening experience. We immediately went about solving this cost-added opportunity.

At first there was a bit of concern about the unwieldy task of significantly shortening this material flow. Many of the machines in the factory had deep foundations made of solid concrete under them. To move these machines would be costly, and if they were not moved to the right spot, it might be money wasted for nothing. It seemed very risky. The answer was to devour this giant project one bite at a time. We started with forks. This seemed simple enough and a great place to start.

Prior to this project launch, it took several weeks to get forks from raw steel to assembly. There were several operations that required a long string of travel

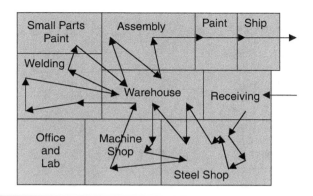

Figure 11.5. Functional Manufacturing Flow Layout.

and waste. The resulting project was a horseshoe work cell that took the lead time from weeks to four hours. All forks for a particular assembly line were made in four hours the day prior to the MPS requirement in assembly. Each day, steel would be delivered according to the day's requirements from the distributor located just a short distance away. These requirements were driven from the MPS shared with the supplier. Just think of it — several weeks to four hours! Once the project was completed successfully, it was actually more surprising that it could last weeks than the fact that it could be logically done in four hours. Figure 11.6 illustrates the work cell implemented.

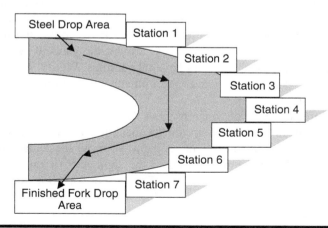

Figure 11.6. Fork Work Cell Example of Process Flow.

DEALING WITH THE WORD "CAN'T"

Some of the lessons learned were also about cultural paradigms. One in particular was a rule that welding "can't be done in the assembly area." The rule was taken for granted because, logically, contamination was not allowed in a clean assembly area. Grime could get into the moving part assemblies and create unwanted friction. What we did not do enough of in the beginning of the project was problem solving! We found later that if the word "can't" was always challenged, the results were often surprising. In this case, we found out through some investigation that welding contamination can be removed successfully at this process source, so that little to no contamination is created.

"Can't" is a funny word. The underpinnings of the word "can't" are divided into three categories. Notice that the odds are with you — 66 percent in your favor that the project is possible:

1. **Incompetence** — Often, this is simply lack of knowledge. It is easy to discover incompetence daily on very new topics. It is not a terminal illness, by the way, as long as it is recognized quickly.
2. **Laziness** — Simply the lack of energy to go after something. This can happen because there is no confidence for success. It often is not called laziness; it is more often called lack of desire. Call it what you want. This is probably a worse problem to deal with than incompetence and may seem harder to recognize by yourself.
3. **The laws of physics (that really do make something impossible)** — What the word "can't" was invented for. Some things just cannot be accomplished. There are huge problems with this rule because the impossible is done every day somewhere. In the last few years, many developments in technology have made the impossible happen. Many things within the "laws of physics limiting progress" at one time are now taken for granted. It is very difficult to understand what is not possible. To make matters more confusing, something that is impossible today might be possible tomorrow. The hardest part about my job as a consultant is to understand what is impossible because I have seen "the impossible" done so many times. I always try to start from the position that it might be possible.

The three-part descriptors for "can't" may seem a bit harsh. These descriptions are designed to get your attention on an important message. Do not assume goals "can't" be done before a lot of poking, prodding, and testing. Make sure the laws of physics are in play.

SETUP/CHANGEOVER REDUCTION

The next topic that is frequently a major opportunity in almost every manufacturing facility is setup reduction. Also referred to as changeover reduction, this is a great area of focus with almost unending potential. Facilities with machines that convert product are only making money when the machines are actually converting needed product. Any time machines with customer-driven work to do are sitting idle, lost-opportunity cost is being incurred. If a machine is able to convert products in one hour that equate to profit when shipped, the machine is actually earning a profit for the company at that rate per hour. When the machine is in changeover and idled, no profit is produced; in fact, quite the opposite is going on: costs are being incurred.

There is more to the story, as APICS-trained people know. If a machine is not changed over, often there is a natural requirement for additional buffer inventory to "fund" the longer run times. Small-lot orders or even one-piece orders are the direction most high-performance businesses find to be the most profitable, taking all costs into consideration. These considered costs include inventory charges as well as changeover times amortized on their typical lot size. So now the story becomes more complicated.

Lean thinking and techniques focus on the elimination of waste. Changeover time is waste. One method to minimize this setup time that has been very successful in literally thousands of businesses worldwide is single minute exchange of die (SMED). Developed by Toyota and credited to Dr. Shigeo Shingo, the SMED approach is an easily learned and communicated process methodology for many activities including machine changeover and even preventative maintenance.

SMED starts with an understanding that all activity originates from two distinct processes: internal and external activity. To best explain the differences, think of a simple drill press such as many hobbyist carpenters have in their shops. If the drill press operator is drilling three pieces of part A, which requires a three-quarter-inch hole, and then has to change the setup over to drill three pieces of part B, which requires a one-half-inch hole, there are some activities that would normally be planned and executed. Note that the following list of activities may not be in the best sequence. It is just a list of activities that need to happen in some order.

1. The one-half-inch drill bit is sharpened.
2. The one-half-inch drill bit is brought to the machine.
3. B parts are brought to the drill press.
4. Completed A parts are moved on to the shipping department.
5. The three-quarter-inch drill bit is sharpened for the next use.

6. The three-quarter-inch bit is put away in its storage bin.
7. The three-quarter-inch bit is removed from the drill press.
8. The one-half-inch drill bit is inserted into the drill press.

Using Dr. Shingo's methodology, each of these steps required by the process falls into either internal or external activity. Internal activity is activity that happens with the machine shut off. External activity happens while the machine is still running. Any activity that can be external allows the machine to continue earning money. For the list of activities for the drill press, in a sample case where the machine operator shut the machine down before doing any of the activities, these activities would be considered internal activities. In this methodology, internal activities are to be avoided or minimized if at all possible to keep the machine running. Examining further, we could imagine that the only steps that really require the machine to be stopped are steps 7 and 8. All the others could be done with the machine still running. Granted, there might have to be some help, but it could be done. This is the idea behind SMED. Make sure the machine is kept running, but allow for a lot of changeovers.

Using SMED keeps costs down and is easy to both understand and administer. SMED is normally implemented using employee teams after initial education and training have been completed. Results can be surprisingly significant. In one organization, a changeover in a plastics business took over twenty-four hours to do. In this application, there was a machine line that was probably twelve feet wide and sixty feet long. The sequence of operations in this line was extruding of plastic sheet, vacuum forming hundreds of parts in a multicavity mold, blanking the parts from the sheets after forming, and stacking the parts for packaging into corrugated boxes. Overhead cranes were used for hoisting these huge tools in and out of the process during changeover. Changeovers were long, but people working on the line were quick to tell of how this changeover had been reduced several times over the years to the existing time standard of *twenty-four hours*!

A team of workers from all shifts was assembled and charged with reducing the setup by 75 percent. After only a short period of time, about 90 days, the result using SMED was a reduced changeover of only 1.6 hours (from 24!). This was an enormous help in reducing inventory. In this case, the company manufactured products for consumption in restaurants. Flexibility and responsiveness were key customer requirements. As one would expect, the company's biggest and best customers did not want to carry inventory. Flexibility meant that the company making the plastic items did not need as much inventory. It could change over more rapidly and replenish quickly, which gave it a big advantage in the marketplace.

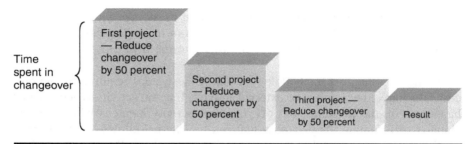

Figure 11.7. Cutting Changeover Time by 50 Percent Then 50 Percent Then 50 Percent.

Dr. Shingo used a philosophy that has since been adopted by many firms around the world. His approach was to start every SMED project by establishing an objective, such as to reduce changeover time by 50 percent, the first time through on any particular setup regardless of opinion regarding the potential on this stockkeeping unit. Once this is successfully accomplished, the next assigned team will have the same goal: to reduce the setup by (another) 50 percent. By the time the third team is chartered to reduce an additional 50 percent, the setup is more like 10 percent of the original at the time the projects were started (Figure 11.7). This approach is a good one that many companies have found to be helpful. By setting the objective high, the results are always improved.

CUSTOMER-FOCUSED SIX SIGMA QUALITY

Both lean and Six Sigma fit so well into this piece of Class A that it is unthinkable not to include these methodologies and training in the implementation plans. No business works strictly on Class A implementation without utilizing lean and problem-solving tactics. In fact, it is difficult to separate Class A from lean and Six Sigma; the three are so integrated in high-performance organizations. These three process focus areas are not sequential in entirety, but instead are concurrent. These implementation methodologies may, however, start sequentially, as seen in Figure 11.8. This illustration shows that the focus may be specific in each case, but the activities are concurrent much of the time. The focus areas migrate as the maturity of the organization develops and the problem-solving tools become second nature.

> The focus areas migrate as the maturity of the organization develops and the problem-solving tools become second nature.

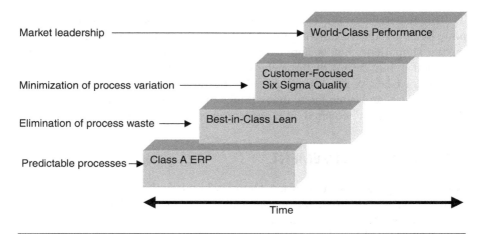

Market leadership ⟶ World-Class Performance

Minimization of process variation ⟶ Customer-Focused Six Sigma Quality

Elimination of process waste ⟶ Best-in-Class Lean

Predictable processes ⟶ Class A ERP

Time

Figure 11.8. Journey to World-Class Performance.

The main thrust of Class A is process ownership, accountability, and establishment of a management system. As lean is introduced into the thinking, the elimination of waste becomes a bigger priority and focus utilizing the management system established in the Class A implementation. As process improvement matures and the low-hanging fruit becomes less and less obvious, Six Sigma methods can kick in and move the organization to the next level of performance. In this chapter, the emphasis will remain on the shop floor regarding the topics of lean and Six Sigma even though there is much more opportunity in most businesses. Chapter 19 will take a harder look at integration of lean and Six Sigma throughout the organization while implementing a Class A ERP process.

On the shop floor, there are limitless opportunities to incorporate lean thinking into the everyday process. Some examples include:

- Kanban (pull systems) on component inventory
- Suppliers actually filling the bins on-site in assembly
- Elimination of inventories wherever possible
- Shortening changeover times
- Improving cost standards
- Elimination of material movement in the factory
- Elimination of levels in the bill of material
- Elimination of steps in the process
- Shortening new product introduction implementation time
- Ergonomics

This list could go on for several pages, but the message should be clear by now. Lean is required thinking in a Class A business. It has special application on the shop floor, but is applicable in every corner of the business. Setup reduction is but the tip of the iceberg. Employee involvement is another key element of Class A shop floor control. After all, the employees are closest to the real issues and often know many things management does not realize when the real facts are uncovered. We need to engage their minds.

EMPLOYEE INVOLVEMENT

Not very long ago, manufacturing facilities hired workers to do manual tasks only. In today's environment, markets are so competitive that it is no longer feasible to waste any resources an organization has. One of the assets that requires little additional cost to harvest is human brainpower. When many of the people who work in the factory leave the plant at the end of the day, they are Little League coaches and Brownie leaders, treasurers of the church, presidents of the rod and gun club, and organizers of unending lists of activities. Historically hired for their backs and strength only, today they are needed for their problem-solving capabilities as well. I recently worked with a facility where this potential was just being realized. In this organization, there is a major focus on problem-solving teams made up of shop floor workers. This is a new concept in the facility and one that, at first, required coaching and training of the line management. Traditionally, this organization did not encourage people to do anything but the work called for in their job descriptions. Problem solving was not normally written into these documents. Little time was allowed for training or problem solving. The transition is not only helping the company to meet its goals, but is also helping to increase the cooperative nature of the labor union in this facility. This is good because relations have sometimes been strained in this company. People want to be part of the decisions and have an influence in reform.

> Business is not a democracy. It does not work well as a democracy. It requires direction and a willingness from management to lead.

Keeping people involved starts with the understanding of company goals. Business is not a democracy. It does not work well as a democracy. It requires direction and a willingness from management to lead.

With this as a starting point, it becomes easy to have employees engaged. As discussed in Chapter 3, once the business imperatives are well communicated, the implementation of these objectives can be initiated. Class A ERP criteria require continuous improvement teams to be involved in the following and more:

1. Implementation of new products in production
2. Cycle time reduction for line speed and/or process
3. Accuracy of inventory transactions
4. Elimination of unnecessary operations
5. Reduction of inventory movement
6. Setup reduction
7. Workplace organization and housekeeping (5-S)

Having teams involved makes the solutions more acceptable to the masses and, when properly facilitated and led, ensures better solutions. This can include projects that are not popular among the workers when initially started. One team in Scotland was working in a remote building just a few miles from the main plant. Improvements in process design at the main facility had allowed this satellite operation to be considered for transition back into the main plant. Initially, it was not a popular move for the employees who worked at the satellite facility. They enjoyed autonomy and worked at a pace that allowed them to always be ahead of the main plant to some degree in terms of schedule and stock. Moving this operation into the main assembly plant meant that these employees would receive much more influence from management and the support infrastructure. This obviously had pluses and minuses in their view. To top it off, the spot they were moving into was less than 50 percent the size of the existing space where they were presently working! This would have been a recipe for disaster had it not been for employee team involvement. The team was facilitated by a Six Sigma master black belt, Gardiner Arthur. He created a team project to have the main players in the facility from each work area resolve the problem of how to fit into this new space effectively. By having people from the area do all of the problem solving for the solution, the emphasis and associated energy were diverted from the change decision itself and focused on how to make the decision work well. It was a good decision and the solution was made equally desirable by the team. The final solution allowed the elimination of a couple of days worth of buffer inventory between operations and thousands of dollars in savings. There are myriad examples of similar successes where employees were used to find the best solutions.

TEAM PLANNING AND MANAGEMENT

Good team management is imperative for maximum success and payback. Along with project management, covered in a separate chapter, team planning and management is helpful in aligning resources properly. It is very fitting for a topic labeled enterprise *resource* planning.

Teams are most efficiently managed when the membership is kept to a minimum. Depending on the complexity of the project, most teams can be run effectively with less than six members, often with four being an optimum number. Teams such as a setup reduction group can be effective with even fewer team members. Roles of the team members and support personnel are important and the right starting point for this discussion.

Sponsor Role

The sponsor of a team is the top manager who wants the task completed. A new product introduction project might be sponsored by the VP of operations. A setup reduction project, however, may be sponsored by the production manager responsible for the machine shop. The sponsor's job is to keep clarity around the objective, help with initial input on the goal, and sign off on the project scope written by the team. The sponsor's job ultimately is the success of the team (Figure 11.9). This means that if there are barriers to success, the sponsor may have to "carry water" for the team. This is not to say the sponsor is going to do the team members' work. Examples of needs supplied by the sponsor might include aid such as:

- Picking the right members for the team
- Choosing a qualified leader for the team

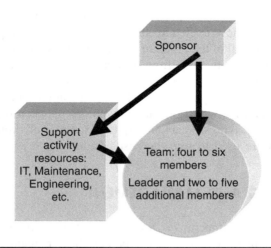

Figure 11.9. Team Structure.

- Arranging required engineering support
- Authorizing monies required for specific solutions
- Arranging for problem-solving training
- Changing a priority within information technology for quicker support
- Dealing with disruptive behavior on the team

Sponsors need to stay current with team progress. This usually means having regularly scheduled review meetings and even attending the team meetings occasionally. If the project is truly important to the sponsor, failure will not be an option.

Team Leader

In mature Six Sigma environments, qualified team leaders are usually referred to as black belts. The role of this important position is to direct the day-to-day activities of the team. It is also their job to facilitate the problem-solving methodology and decision-making process. Basically, the team leader is the person accountable or "on the hook" for the team's success. The sponsor is responsible, but the team leader is both responsible and accountable. The team leader might be involved in activities such as:

- Scheduling the team meetings
- Having an agenda prepared for the meetings
- Making tie-breaker decisions when necessary
- Facilitating problem-solving exercises
- Communicating progress reports to the sponsor
- Getting support activities scheduled

Obviously, the team leader has a lot of responsibility for the team. This person needs to be well qualified and is normally handpicked by the sponsor. It works best when the team leader has training in both decision funnels and problem solving and is well respected by the team members. It works best if the team is smaller, but in some instances, more members can work together successfully if the team members are seasoned to team-based problem solving.

Time allotment is also a question many times. In most successful team-based efforts where teams are populated with workers or people from the actual processes being studied, one to three hours per week per member is the norm. In some instances, this time is budgeted by the team. If, for example, the team has two hours a week to meet and do team-based activity, and there are four team members including the leader, there would be eight man-hours available and authorized for team activity per week. Sometimes teams use the time more wisely by keeping meetings to a minimum (say one hour) and using the delta (in this case, four man-hours) to get assignments completed. One team member may use the entire four hours left that week to do a specific study, to gather information, to meet with a support person, etc. If the meeting were shortened even more to forty-five minutes, more time would be available for team assignments. This is a helpful and sometimes overlooked method to get more done in teams with limited available time.

KAIZEN EVENTS ON THE FACTORY FLOOR

Kaizen events are one of many lean Japanese introductions to American manufacturing culture. Toyota, again, is usually credited with the first use of the word in this process improvement context. According to Masaaki Imai, author of the book entitled *Kaizen* (McGraw-Hill, 1986), kaizen means "gradual unending improvement, doing 'little things' better; setting and achieving even higher targets." In this application, from the perspective of the shop floor, it means that fast bursts of energy and focus are pursued on specific processes to support continuous improvement. These focused events are a very efficient form of continuous improvement. Kaizen events normally are performed using one of two popular approaches.

One approach, the twenty-four-hour version, is all about a short burst of activity meant to get as much as possible done on a specific process element in one day. The other popular approach, the one-week version, is longer, but still very much focused on accomplishing as much in the week as possible. These high-energy approaches have proven to be very helpful in getting more process improvement accomplished in the short run as compared to working this activity concurrently with other normal daily requirements. As is widely understood, "working it concurrently" sometimes is better in intention than implementation. Both the twenty-four-hour and the week-long versions have their place in Class A plans and execution.

The twenty-four-hour kaizen is frequently done at the request of a sponsor and as support to a business imperative. One of the most common twenty-four-hour kaizen events is used to reduce setup time or process changeover. In a typical project, the sponsor may or may not pick the particular changeover to focus on. If not, the activities might go like this:

1. Business imperatives call for increased flexibility and reduced inventory
2. Sponsor engages setup reduction team to work setup reduction in support of these business imperatives

3. Team chooses specific setup to focus improvements on
4. Sponsor provides training resources for kaizen training and training in SMED techniques
5. Team leader facilitates a project "plan of attack" with the team and the team then discusses the strategy to achieve this goal
6. Team chooses date for kaizen event
7. Setup is filmed in preparation for the kaizen
8. Master scheduler schedules the slot into the weekly plan for the kaizen event
9. Kaizen is held
    a. Team watches video of setup
    b. Internal and external setup activities are identified by the team
    c. Internal activity is recognized for conversion to external activity by the team
    d. Agreements are made for resulting actions to be completed by the team
    e. Actions that can be completed in the single day are assigned to team members
    f. Actions are completed

In a typical kaizen as described, the particular setup might be reduced as much as 50 to 60 percent or more in that single day of effort. Follow-up events may leverage lessons learned, applying these improvements to other similar setups. In many situations, documentation of changes would follow the kaizen event as an assignment to a team member. It is also common with this type of kaizen to orchestrate another event for the same setup in just a couple of weeks. The goal on this second go-around would be the same 50 percent reduction of time elapsed during the setup. The timing is important because everyone at this point has the setup elements fresh in their minds from the recent improvement work.

Other Class A ERP kaizen events could center on cycle time reduction or waste elimination in a process. Data accuracy would also be a very appropriate Class A–driven kaizen. There is almost no end to the possibilities. The limits are governed by the scope of the business imperatives. Since cost reduction is almost always in the objectives, kaizen events are always helpful when administered appropriately.

Week-long kaizen events are the same process, but are centered on more complex opportunities. Some ideas that have been commonly used in Class A businesses are firs- time quality improvements and redesign of a production line for elimination of material movement (including movement of equipment, 5-S focused efforts, etc.). There could be variations of kaizen events between one day and one week. It just depends on the complexity of the task at hand — how complex it is and how long it will take to make some significant progress given full attention and a reasonable amount of trained resource. The steps for a normal kaizen event are (after training of participants and assignment of team):

1. Identify area of waste opportunity
2. List waste
3. Prioritize waste by time to accomplish and effect on the business imperative
4. Determine methods to eliminate the waste
5. Take action
6. Measure results
7. Standardize
8. Celebrate and start over

As you can see, the process is a variation of the Deming wheel. Many of the more effective management tools are based on the Deming principle of Can-Do-Check-Act. It is always the simple principles that endure! The kaizen process format also has a lot of commonality with the Six Sigma process of DMAIC (Define, Measure, Analyze, Improve, Control). More detail on DMAIC and Six Sigma will follow later.

AREAS OF WASTE

There are always many examples of waste in manufacturing and/or distribution processes. There is also no end to process variation, only levels of difficulty recognizing it. Much of the variation in most companies is still pretty obvious. These areas of waste represent the opportunities to fuel continuous improvement throughout the facility. Recognizing them is the obvious first step. Some of these areas include:

1. Unnecessary material movement
2. Unnecessary worker movement
3. Wasted space
4. Wasted time
5. Wasted material
   a. Scrap
   b. Rework
   c. Lost material
   d. Damaged material

     e. Mistakenly ordered material

     f. Unnecessary material ordered or manufactured

6. Wasted effort (poor quality)

     a. Work quality

     b. Material quality

     c. Process design quality

     d. Standards

7. Wasted knowledge or information

8. Wasted creativity

These areas of waste represent the opportunities to generate a list of project needs for your shop. Every plant and distribution facility has them; it is just a matter of pursuing them. It is widely recognized that a great majority of the money spent in managing and running an operation does not go directly to cost-added activities. Too much goes to the list of wasted resources listed above. Often by starting from this list, teams can generate ideas quickly to move the needle on the ratio of value-add versus cost-add activities in the facility.

5-S HOUSEKEEPING AND WORKPLACE ORGANIZATION

No discussion about shop floor control (SFC) would be complete without the topic of housekeeping (5-S) and workplace organization. 5-S also originated in Japan with the Toyota organization. All of the 5-Ss actually are Japanese words focused on neatness and efficient use of process-related storage. The translation for 5-S is:

1. **Seiri (or Sifting)** — "Sift" out equipment and tools not normally used in the process and get them out of the area.
2. **Seiton (or Sorting)** — Make sure that there is a place for everything and that everything is in the designated storage place.
3. **Seiso (or Sweeping)** — Sweep often and well. Sweep behind and under things, for example conveyors, etc. Have extremely high standards for cleanliness. Do not put up with black skid marks on the warehouse floor, have employees put trash in proper receptacles, etc.
4. **Seiketsu (or Standardizing)** — Practice discipline. Document the expectations. Anything that is to be maintained needs to be written down for compliance verification. Any process requirement that is not documented cannot be audited. Documentation does not have to be taken to an extreme level, but is a necessary tool unfortunately.

| Department Audited _____ | Date _____ | |
| Audited by: _____ | | |

Topic	Expectation	Rating
Visual schedule boards	Updated and current	_____
Organization of tools	Tools not in use put away, places for tools	_____
Organization of equipment	Tools in good condition, no oil leaks	_____
PMs done and documented	PM sheets visible and current	_____
Clutter	Area neat of clutter	_____
Documentation	SOPs available for processes	_____
Aisles	Clear, clean, swept, no tire marks	_____
Lines of demarcation	Lines painted clearly defining drop areas and work cell	_____
Location identification	Areas are well marked including storage locations	_____
Cleaning equipment	Available and neatly stored (brooms, garbage cans, etc.)	_____
Shelving and racks	None in area that are not necessary, are clean and neat	_____
Safety equipment access	Fire extinguishers, eye wash stations, etc. easily accessible	_____

Safety Violations   **None allowable**

                              Total Score _____

Figure 11.10. 5-S Sample Measurement Tool.

5. **Shitsuke (or Sustaining)** — A management system must be in place for sustainability. A weekly review or similar follow-up is often effective. Some organizations create competitive fun by offering incentives to be the best. One organization offered extra vacation time if any department could be the best in the company for three weeks in a row.

Using these approaches to make the operation more efficient is a logical management technique and criterion for Class A performance. An example of a measurement tool that incorporates these organization criteria is illustrated in Figure 11.10. This tool is best if used throughout the organization, both office and factory (or warehouse) floor.

SIMPLE PROBLEM-SOLVING TOOLS

Simple problem-solving tools are a necessary part of good shop floor management (see also Chapter 13). These tools would include:

- **5-why diagramming** — Simple five-question format where each suspected reason is brainstormed and then questioned again and again, level by level, much like peeling an onion.

- **Brainstorming** — A method of making sure all team members are engaged and that all ideas, both good and bad, are captured and later evaluated.
- **Fishbone diagrams** — Giving structure to the reasons can remind team members of areas they might otherwise ignore. A simple fishbone diagram reminds them that there could be input variation opportunity in areas of method (process), manpower (people), material, machine (tools and equipment), and environment (temperature, barometric pressure, altitude above sea level, humidity, etc.).
- **Decision-making funnel** — A common process to prioritize ideas into agreements. Different methods such as voting, weighted voting, high-priority/low-priority accessing, applying criteria, consolidating, etc. are appropriate in different applications.
- **Problem-solving methodology such as Six Sigma DMAIC** — A variation on the Deming Can-Do-Check-Act methodology (see Chapter 13).

Employees need to have education and training on these tools. Exposure alone is not enough. Management needs to "walk the talk" and use the tools just as they are expected to be used on other teams. If these teams are allowed the time and given the training and management support, a continuous improvement culture can flourish. There is no better method of SFC than that.

PROCESS OWNERS FOR SHOP FLOOR CONTROL

Single ownership of processes almost always is the right approach. Having stated that, this is the one category within Class A ERP where there are normally multiple process owners. Even after making the case that accountability needs to be with one single resource, in this instance there is benefit in delegating this accountability. Each supervisor is expected to deliver on the daily schedule. If it is not going to happen and this supervisor knows it in the morning, the schedule should not be committed to by the supervisor. If the schedule is going to be beat, the same thing is true; the supervisor has an obligation to inform the master scheduler of this information. Each supervisor does not have to report this metric to management, however. In some situations, the supervisor has a team-working role and it is difficult to attend management system events such as the weekly performance review. Sometimes the manager with responsibility over the supervisors will report for the group of production teams as a whole. Individual team performance would appropriately be visibly posted in each department. The weekly performance review could also be populated with rotated representation. This will be discussed further in Chapters 12 and 14.

THE METRICS IN SHOP FLOOR CONTROL

The Class A ERP metrics on the shop floor are daily metrics. The MPS is agreed to for the week and frozen for measurement purposes. This main schedule designates the weekly format and metric for consolidated performance, leaving the metric space open for the SFC metric. This daily metric has a different set of opportunities. In every business, shop management needs to plan the day each morning and commit to certain levels of production and specific items/quantities to be completed. At times, this may mean that this commitment has changed since the MPS was agreed to last Friday, but nonetheless, adherence to this revised schedule is no less important. Additionally, in many shops, there can be some leeway within the daily schedule to accomplish the plan. In other words, the master scheduler may say to each department, "Here is what you need to have done *by the end* of the day." In reality, the order of completions within the day may or may not necessarily be all that important. Note that in other businesses, the MPS is to the hour instead of the day. In these hourly scheduled businesses, the order can be very important and the rules basically stay the same; only the time interval of measurement changes. Whichever scenario your business falls into, the common denominator is agreement and measurement to that agreement.

The SFC commitment is matched to the completions that need to be done at the end of the day. If there are forty orders scheduled with the fortieth scheduled to be half done at shift end, the measurement is to the scheduled completions, forty of forty complete. The balance of that last order will get picked up in the measurement the following day. In companies where one-piece orders are the norm, unfinished orders are not usually an issue. Each day, the goals are set for the day linked to the MPS. The master scheduler is involved in this process and updates the MPS as required accordingly. In production operations where process variation is controlled and on limited schedule, "noise" is minimal. Either way, to meet the Class A criteria, the commitment must be made or the schedule adherence evaluated at the end of each day.

Machine Line 4213	Mon	Tue	Wed	Thu	Fri
MPS of record	200	210	211	199	205
Revised MPS (Tuesday)	200	215	228	210	208
Daily commitment	200	215			
Orders completed on time	200	218			
SFC percentage	100%	99%			

If the MPS stayed stable through the rest of the week and machine line 4213 was able to complete the schedule as revised within Class A boundaries, the schedule might look something like this:

Machine Line 4213	Mon	Tue	Wed	Thu	Fri
MPS of record	200	210	211	199	205
Revised MPS (Tuesday)	200	215	228	210	208
Daily commitment	200	215	228	210	208
Orders completed on time	200	218	225	210	207
SFC percentage	100%	99%	99%	100%	99%

What probably happened in this case is the sales department sold more units than was originally planned. This caused the MPS to be revised and the daily schedule to be changed to match the MPS. The MPS of record would show the deviation, allowing the appropriate learning to happen as accuracy is sought for the MPS plan. The SFC metric has a different purpose and focus from the MPS metric. In this case, the business is focusing on being able to plan the *day's* activities and being repeatable in making the commitment. The average daily ability to make a plan in this example would be 99 percent or (100 + 99 + 99 + 100 + 99)/5. This is not a measure of how many shop orders were done in the day they were scheduled; it is a measure of how accurate the plans made each morning are.

It is also acceptable within Class A criteria to do the other scenario (orders on the day scheduled) as long as you are able to update the plan of record each day as the MPS is revised. In this case, the performance for the schedule above would be total scheduled divided by the total missed subtracted from 100 percent or 100 − [(0 + 3 + 3 + 0 + 1)/(200 + 215 + 228 + 210 + 208)] or 100 − (7/1061) or 99 percent.

Each day, the shop floor supervisors or managers commit to the daily schedule and are measured to this commitment. Be aware that it is possible that the same order can be missed more than once in the measurement if orders are scheduled for Monday and are missed and are rescheduled for Tuesday and missed again. In fact, if the scheduling job is done *really poorly,* this can happen over and over. In a daily metric, the expectation is that the schedules will be real. If the same order was inaccurately scheduled multiple times in the week, the misses should show up multiple times.

FIRST-TIME QUALITY

Quality is also a measurement of shop floor effectiveness. The Class A measurement of choice for shop floor quality is the first-time quality (FTQ) metric. This is a main focus later in Six Sigma performance requirements, but for this first wave of effectiveness and process foundation building in Class A, the threshold of acceptability is something short of perfection. In some businesses, this acceptability must be at least 98 percent or 3.6 sigma; in others, 3.2 sigma or 95 percent is an acceptable starting point. The minimum acceptable Class A ERP threshold is 95 percent or 3.2 sigma.

FTQ is measured by understanding the percentage of units that makes it through the process with no interruptions. This includes both valid rejects and units that were inaccurately sidelined that had no defects when checked further. All units that go through completely without exception handling, exception checking, or exception rework and are within acceptable limits as defined by the customer specifications can be considered FTQ units and represent the numerator in the ratio when calculating the measurement. The denominator is the number of units started in the process.

I worked with a factory that manufactured large products. Frequently, the production line, under management direction, would continue even though there were parts missing in the process. This was handled by finishing the units as far as possible and setting them off to the side in a "boneyard" to wait for assembly of the missing part when it was later received. In the first follow-up visit after the initial education, I discovered that "boneyard" units were not being considered as misses in the FTQ. Class A would call this an exception handling and therefore it would be included in the daily misses.

SAFETY

The last area of measurement within the SFC space is safety. In the last few years, due to the importance of effective resource management, the desire for low turnover of employees, continuously higher insurance costs, and regard for employee quality of life, safety has grown exponentially in importance within the Class A ERP criteria. To ignore safety in manufacturing is to ignore a significant contributor to improved employee involvement, morale, quality, cost, and customer service.

Time without recordable accidents is the metric, and the longer the better. One Class A ERP–certified facility had a safety record of more than one million hours without a recordable accident. Phil Hilling, the plant manager, signs his e-mails with the quote "Be safe, it is the right thing to do." I was actually cited in the plant one day for going down the stairs without holding on to the railing. Safety in this plant was important. It was also cost effective. Because long-term safety is an ongoing focus and because one recordable accident can scrap a planned record for safety for the entire plant, Class A ERP criteria require improvement as the minimum threshold of acceptability. Once the company objectives by division are met at minimums of recordable accidents, the record must show continuous improvement. The discussion on metrics continues. In the next chapter and Chapter 16, all the metrics will be summarized and reviewed.

12

CUSTOMER SERVICE

What is more important than the customer? In business, probably little could be of more value-add than customer service, especially in today's competitive markets where service often differentiates one firm from another in the eyes of customers. Given that, it is surprising that customer service is such a misunderstood topic in many businesses. It is also surprising that there are a number of companies that find a large unexplainable gap when comparing their data to the customers', even though they post a 95+ percent service level. What is even more frustrating is that many of these same companies seem to lack interest in reconciling the difference between the two numbers. There is much to be learned from these gaps. In this case, the customer is (almost) always right.

The numbers do not lie if the people who use them are honest — with others and with themselves. One company I worked with had a bonus program that was driven from customer service. The spirit of the metric was exactly what it should be: the fidelity of the first promise to the customer. The issue developed over "acceptable" exceptions, one of which was customer reschedules. Logically, if a customer called and delayed a shipment due to its own reasons, the company did not feel it should have to count those delays against the metric, especially when the bonuses were calculated on these data. If that is entirely how it worked at this company, this paragraph would not have been included in the book. What happened in that company happens in a lot of organizations around the world. The company would often have warning when it was going to miss original schedules. Usually it happened because parts were delayed or scrapped and replacements were pending. This made it difficult to ship products on the promised or required date. When this happened, the customer would be called and, in effect, asked to reschedule the order. Many times, the customer would agree to the new date when told that the original date would not be

Many times, the customer would agree to the new date when told that the original date would not be possible — after all, there was little the customer could do anyway.

possible — after all, there was little the customer could do anyway.

It did not mean the customer was happy. It would have been good to have this handshake with the customer if it happened only occasionally. It happened frequently, however, and the predictable result was that when the customer "agreed" to a reschedule, the customer service metric was adjusted to measure to the latest, adjusted date — after all, *the customer agreed to it!* This is a good example where customer data and supplier data did not match. The real fidelity was closer to 85 percent, a performance level that is unacceptable and one that would not support bonuses as per policy design. Many companies have similar situations and yet choose to ignore the realities. It always feels good to look at favorable numbers, even if bonuses are not attached to the data.

In yet another company that uses several distribution centers in North America, a slightly different twist on the same theme occurs. In this business, the plants determine the inventory levels and reorder points for inventory of finished goods in the distribution centers (DCs). The DCs accept both the replenishment orders and the customer orders as determined by either customers or producing factories. The DCs were the final handshake with the customer, however. They were looked at, right or wrong, as the representative of the manufacturer. They obviously had to measure promise fidelity and did. They measured how many times they were right on their promises. The DC promises encompassed a delivery appointment made with the customer after inventory was available and ready to ship. Misses usually were only because the over-the-road (OTR) truck driver missed the delivery. The posted performance numbers were above 99 percent continuously. The rest of the organization ignored the data, except for sales, which used to point fingers all the time in dismay. Salespeople knew that promises were being missed at a higher rate. Even they did not really understand the metric disconnect. No one did much about it until Class A requirements started to require a double check. Sadly, this is too common a story. Unless management is poking at it and following up with customer data, there is always room for manipulation. Many times, this manipulation is honest and full of good intentions. After all, the measurement from their perspective needs to be more in line with *their controllable* performance. This is exactly where the system often breaks down. The customer service measurement in Class A ERP is not a measurement of the shipping dock. It is also not a measurement of the performance of the OTR drivers. Customer service is the culmination of all of the functions of the organization. This includes rules governing promises, adherence to those rules

The customer service measurement in Class A ERP is not a measurement of the shipping dock. It is also not a measurement of the performance of the OTR drivers. Customer service is the culmination of all of the functions of the organization.

of engagement, execution of the rules, and strategy to make it happen repeatedly. In fact, it is rarely exclusively the shipping dock's fault that shipments are late. The OTR delivery performance is a valid measurement, as is appointment promise from dock to dock, but it should not be confused with a Class A customer service marker.

RULES OF ENGAGEMENT

Without the rules of engagement properly established and understood as well as communicated to both the sales and marketing people and with the customers, there is little chance of predictable and repeatable high performance. Rules of engagement come in many "flavors," but always start with inventory strategy (make to stock, assemble to order, make to order, engineer to order, etc.). By understanding the expectation of lead time and supporting inventory, salespeople know what to expect and to promise. If it takes three weeks to get inventory in position for shipment of a particular product because of special nonstocked purchased components, it is foolish to promise it in less time, regardless of the customer's severe need. In other words, if the company has positioned itself through a shared agreement with sales and operations to have a three-week lead time, it will probably take most of those three weeks to

In high-performance companies, sales is the main input to the handshakes/rules as they are established. Sales gets to determine these rules and understand the costs associated with various alternatives at the same time.

deliver. If it is the wrong lead time, the rule of engagement needs to be changed to meet need. Every change in inventory position has an impact on cost. That is just the way the world is. Until the lead time is changed through supply chain handshakes or design simplification, it is what it is. The sales organization needs to understand this

and carry the banner. In high-performance companies, sales is the main input to the handshakes/rules as they are established. Sales gets to determine these rules and understand the costs associated with various alternatives at the same time.

Unless the demand-side people enforce these rules of engagement, the opportunity is lost and the expediting (and associated costs) begins to rise once again. If the rules are in line with market need determined by the sales department, customer service will result. These rules should be changed any time they are out of alignment with market need and should be reviewed once a quarter at a minimum. The review would include the master scheduler and representatives from the demand side. It is helpful if the master scheduler brings data to support any moving trends.

CUSTOMER SERVICE EXCEPTIONS

Whenever there are rules, there are also exceptions to those rules. It will happen, and no one should be surprised. Rules of engagement are determined to establish minimal inventory and still meet market demands. When exceptions are made, costs are affected. In many high-performance organizations, exceptions are only made in severe situations where the customer has an especially critical need. Since this is a judgment call, it needs to be a *little* difficult to get the required waiver for an exception. In one organization, for an exception to be authorized, both the VP of sales and the VP of operations had to approve an order to expedite.

It is interesting that in the many organizations I have worked with, when companies get serious about the internal handshake and rules of engagement, the exceptions decrease significantly. Too many times, the exceptions are easily decided with little understanding of the costs associated with them. When confronted with the dollar impact, many of these exceptions seem to go away. It is easy to make promises, especially when somebody else has to deliver them. On the other side, it is imperative to keep promises once they are made.

> It is easy to make promises, especially when somebody else has to deliver them.

DELIVER ON TIME OR SHIP ON TIME

In many organizations, there are many metrics to designate customer service success, including ship on time and deliver on time. Both of these are metrics that are helpful in specific situations. If OTR delivery is inconsistent, deliver on time is an important diagnostic measure to add to the shipping performance. If loading the trailers for shipment takes time and is inconsistent, there could be an obvious need to measure the process. In Class A ERP, the focus for customer service is shipping on time as the threshold of acceptability. The minimum goal is 95 percent. Delivery to the requested date is the next notch up from Class A.

Some companies are moving away from promises and toward customer requests as the goal to be measured to. In Class A, the focus is on the basics. Step one is to make promises and keep them. The threshold for Class A certification is 95 percent promise adherence to the original date. There is a lot to be learned from this stability especially in organizations that have customer service gaps between need and proven performance. This learning can get lost if customer request dates are the only objective in the initial customer service focus. Remember that if the rules of engagement are the correct ones and these rules support competitive advantage, the capabilities should be in line with or even better than most of the customer requests. When 95 percent promise fidelity is met, ratchet the metric to customer request fidelity.

PROCESS OWNERSHIP FOR CUSTOMER SERVICE

Customer service belongs to everybody in the business. Few can say their part has no impact. For this reason, there is no obvious process owner for customer service. Instead, there is some flexibility on this topic. The criteria for this ownership are willingness to get the homework completed and understanding of the supporting data. In one organization, the general manager's administrative assistant was the process owner. She was very appropriate as the process owner and one of the best at this because she took a high level of interest in getting the data and understanding the root causes for customer service misses. She also carried the clout of the general manager's office when she needed it and was not afraid to use it to get to the root cause as necessary. In other organizations, I have seen success from process ownership in all of the following areas of responsibility:

- Secretary to the president
- Master scheduler
- Production manager
- Manager of customer service
- Order management manager
- Plant manager
- Materials manager
- Procurement manager

The key to the success of this process owner is to have the support of the organization in getting data and affecting change required as indicated by root cause. The process owner is required to report performance data; trends analysis charts; Pareto charts of root cause; and actions, names, and dates to achieve Class A performance or better. Anyone in the business who is ready, willing, and capable to carry this load can be the right person.

THE METRIC FOR CUSTOMER SERVICE

The measurement for Class A customer service, as a minimum, is the fidelity of original ship promises being met. Obviously, inventory strategy makes a difference. In the make-to-stock environment, there is no promise other than to have it in stock when the customer calls. The Class A ERP requirement minimum is to ship 95 percent of orders complete within twenty-four hours. The promise of make-to-stock strategy is to have product ready to ship immediately. In make to order, the promise at the time of order receipt is the plan of record and, again, the threshold of acceptability is 95 percent shipping on time to that original promise.

Some customers will choose to pick up the product with their own vehicles or will arrange transportation with a third party. In these cases, the product has to be ready for pickup, with notification to the carrier or customer on the date promised. For Class A compliance, this exception does not apply to shipments where the customer has specified the carrier, but the responsibility to arrange and schedule the carrier and ship the product still lies with the supplier. In this case, even though the customer has requested a specific carrier, the metric is "ship on time to the promise." If the carrier does not show up, it is still the supplier's problem to deal with and eliminate reoccurrence.

$$\frac{\text{Complete orders shipped on or before the original promise date}}{\text{All orders scheduled for that period}} = \text{Percent ship on time}$$

SHIPPING AHEAD OF THE PROMISE DATE

When product ships ahead of the promise, if acceptable to the customer, the shipment is considered to be on time. Many customers today will not allow shipments to be made early (especially if they are adhering to Class A rules). In these instances, early shipments would be misses, not hits, in the metric. The general rule is if the shipment can be converted to cash (in other words, accounts receivable) and the customer accepts the early shipment, the transaction is considered in compliance with the Class A criteria.

13

CLASS A ERP PROJECT MANAGEMENT/SIX SIGMA PROCESS INTEGRATION

Class A is all about continuous improvement. As was stated earlier in this book, Class A ERP disciplines are the foundation for good performance. Class A performance in itself is not the end-all or state of perfection. It readies every organization for continuous improvement by providing a management system that requires accountability and process ownership and trains the organization to step up to the requirement of continuous improvement. There is always much more work to be done.

The lean process has been mentioned several times in this Class A ERP discussion, especially within Chapter 11. Six Sigma has also been mentioned, but has not received the right amount of attention yet. In this chapter, we will rectify that situation by showing that Six Sigma, and more specifically project management, is necessary even at the Class A level of performance. Class A's basic building block is shared with both lean and Six Sigma (Figure 13.1). That common denominator is continuous improvement.

Continuous improvement requires employees to be engaged and it requires project management, especially as the

> Class A's basic building block is shared with both lean and Six Sigma. That common denominator is continuous improvement.

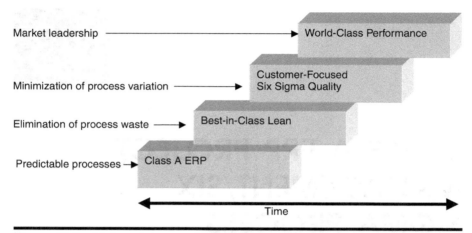

Figure 13.1.   Vision of Journey to World-Class Performance.

low-hanging fruit is eliminated and the tougher tasks start to come to the surface. In many organizations, this is the case. The easy stuff was done years ago. There is no way around it at that point; project management is a requirement.

CLASS A PROJECT MANAGEMENT

Managing projects in a Class A environment is no different than in Six Sigma when done correctly. Under my experiences of success using simplification, I include some of the easier tools in Class A and leave the more progressive statistical tools to the later phases of improvement, Six Sigma. The biggest reason is for clarity only. Six Sigma advocates who have developed their passion while working only Six Sigma and who do not acknowledge a Class A prerequisite step would call all of it Six Sigma. Call it whatever you want; it is project management and it is important. I have been intimately involved in all of it, and here is how I have come to segregate project management.

Class A ERP project management is about setting up the infrastructure for accountability. Like any process ownership within Class A, there needs to be clear project ownership expected and consistently delivered on actions tied to goals. A regular, predictable review process is necessary for accountability to flourish. While I was general manager at Raymond's aftermarket division, a formal project review was held every Friday from 9 A.M. until noon. My administrative assistant set up the schedules, and project updates were reviewed with the same reporting requirements every week. This gave people a chance to show off their accomplishments and gave management a chance to coach and recognize good behavior. This is the spirit of Class A. Our prioritization process was driven from our business imperatives, and assignments were made as resources were available. The secret was to keep project capacity in line with need and not to overassign expectations so as to cause failures.

Class A criteria require that projects be aligned with the business imperatives. Minimally, this means that there is a clear linkage between the projects assigned and planned and the company or division objectives. For example, if lower inventory was a business imperative, a project scoped to reduce setup time on a machine line would have clear linkage to the overall inventory reduction objective. The Class A criterion in this space is the project review by management. Class A uncharacteristically leaves the door open in this category. Normally, most Class A management system events are pretty detailed in the criteria, some even suggesting which day of the week. Project management is a little different. It can differ widely depending on a company's appetite for projects, the resources available, and the criticality of change. Obviously, the need to review projects is the same in every facility, but since the number of projects can vary, project review can vary as well. When I left The Raymond Corporation in 1994, there were 113 projects empowered in the division. Every project could not be reviewed every week, so assignments were made based on priority and frequency of review required. The process worked well. In a Sweetheart Cup plant, the review was held once a month and was done over lunch with the plant manager. Any way it works for you, the basic requirement of Class A criteria is to have clear accountability and a review process that is predictable in both interval and reporting requirements.

A common reporting format for reviewing projects should be adopted in a Class A organization. Microsoft® Project is a popular software tool for managing projects and it is reasonably easy to use. There are others that are equally useful. These tools create a Gantt chart as well as help to define the critical path for project accomplishment. Each project manager in a Class A environment will come into the review with the project status shown in the form of a Gantt chart, finished tasks, tasks being worked, and tasks scheduled or planned. A project schedule accuracy measurement should be reported at each review, as well as any roadblocks to schedule integrity, along with expected actions and results to remove these barriers.

LEAN-FOCUSED PROJECT MANAGEMENT

There is no real difference between the certification requirements for Class A project management and the requirements for best-in-class lean project management; the only difference is in project focus. Whereas the first steps usually associated with Class A ERP are focused on accuracy of data and schedules and additionally on inventory reduction, lean projects more typically focus on projects that reduce cycle time and waste. Both are obviously important, and one focus is not necessarily a prerequisite to the other, not in project management.

Projects in the lean space frequently use process mapping as the main tool, with design of experiment following closely. Class A tools generally are the beginning tools and often do not go beyond simple tools such as brainstorming, Pareto charts, 5-why diagramming, and fishbone analysis. In Class A, the emphasis is on establishing the cultural underpinnings of continuous improvement. Some process mapping is done during the Class A focus, but often it is limited to logical, process-flow-type diagrams. It simply depends on the resource maturity and experience in process mapping. In the lean space, there is ample opportunity to use various process mapping tools such as:

- **Value organization alignment mapping** — A mapping exercise to track information or decision making through an organizational chart (Figure 13.2).
- **Time value mapping** — A methodology used to map both activity and time duration of a process. This is a unique approach in that time value becomes visible in the diagram for each step. Also, by placing value-add process elements below the time line and cost-add activities above the line, visibility becomes a driver for change (Figure 13.3).
- **Swim lane flow charts** — Show activities separated by "lanes" of functions in the process map (Figure 13.4). These are helpful for tracking information or material movement through different parts of the organization.
- **Physical process maps** — Maps of building layouts showing material flow, people movement, or information flow mapped on the blueprint (Figure 13.5).
- **Logical process flow map** — The most common; simple process maps depicting all activities in a line with decision points and various alternative routes shown with specific shapes (Figure 13.6).

Lean project management has a lot of overlap with both Class A and Six Sigma project management. It is really unnecessary to differentiate the processes.

Lean project management has a lot of overlap with both Class A and Six Sigma project management. It is really unnecessary to differentiate the processes. What the real value-add comes from is robust and organized project management. It does not matter what you call it.

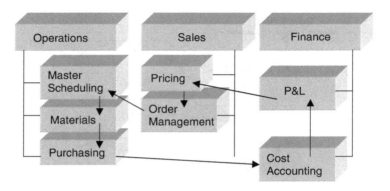

Figure 13.2. Value Organization Alignment Map.

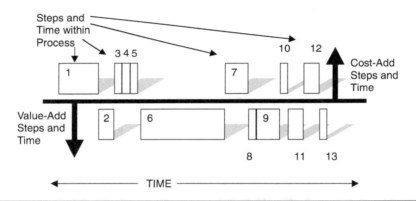

Figure 13.3. Time Value Map.

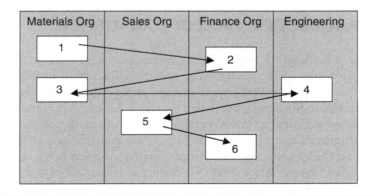

Figure 13.4. Swim Lane Flow Chart.

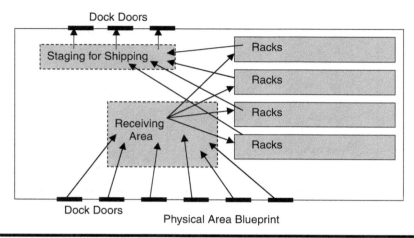

Figure 13.5.   Physical Process Map.

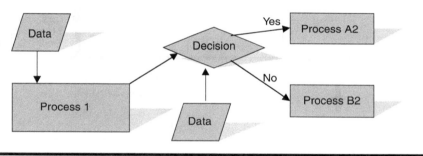

Figure 13.6.   Logical Process Flow Map.

SIX SIGMA–FOCUSED PROJECT MANAGEMENT

Six Sigma is the king of project management processes. What makes this so is the robust nature of the methodology and disciplines built into the process. Like the Class A logical first step in process improvement, discipline is especially important in Six Sigma, but unlike the Class A focus, Six Sigma has much more focus on statistical tools and certification of both skills and project effectiveness of individuals in the methodology. These certifications are referred to in martial art "belt" terms such as (but not limited to) green belts, brown belts, black belts, and master black belts. Six Sigma also uses all of the tools mentioned in Class A and lean project management such as brainstorming and process mapping.

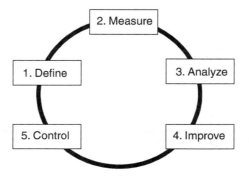

Figure 13.7. Six Sigma DMAIC Model.

Problem solving is a main practice within the Six Sigma process. The Six Sigma practice has been widely recognized for the problem-solving-wheel-type process (Figure 13.7). Although there are different versions out there, the most popular Six Sigma approach is clearly DMAIC (Define, Measure, Analyze, Improve, Control). Motorola and GE are often given credit for the popularity of that particular model. DMAIC is a specific decision-making process that is based on the Deming Can-Do-Check-Act wheel.

DMAIC is simply a structure to remind the problem solvers to follow good practice. It reminds all of us to make sure that the project is well scoped, underlying data are gathered, good analysis is executed, actions are taken, and the solution is hard tooled through documentation or process management systems. The steps are defined as follows:

- **Define** — The first step in launching a project is to define the scope and expectations of the project. This starting point can make or break the project's effectiveness. The project needs to be aligned with company objectives, aligned with available skills and resources, and timed appropriately. Projects with longer time frames are often less efficient. Breaking longer projects into smaller pieces can often result in higher productivity over time. A common mistake in scoping a project is making the objective too broad. Words like "improve," "increase," or "reduce" are red flags unless followed by some specific requirements, such as reduce by 50 percent.

> Breaking longer projects into smaller pieces can often result in higher productivity over time.

- **Measure** — Data are what makes problem solving easier and effective. The second point in Six Sigma problem solving is the reminder to make

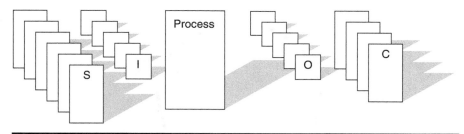

Figure 13.8. SIPOC (Suppliers, Inputs, Process, Outputs, Customers).

sure that all solutions are driven by facts rather than opinions. Gather data from various sources. A popular tool to use in this phase is SIPOC (suppliers, inputs, process, outputs, customers) (Figure 13.8). Keep in mind that the objective is to get to the data and facts. As an old boss of mine used to say, there is "math" behind every process. During this step, the right thing to do is to get to that "math" or facts.

- **Analyze** — This step in the process is where good problem-solving tools are put to use. Statistical tools such as Mini-Tab are frequently utilized. These handy tools are used to chart data and illustrate relationships between data, allowing views from different perspectives. These Six Sigma tools are also helpful in setting up control charts. Control charts make process analysis easier by documenting process variation in relation to expected limits. These tools are more sophisticated and are normally used by more experienced Six Sigma advocates. They generally require a solid understanding of statistics. In many projects, it is not necessary to use sophisticated tools. Many times, simply tracking processes through simple data collection, establishing metrics, and analyzing inputs will quickly lead to sources of variation that can be dealt with in the next step.

- **Improve** — This step is the reminder that unless the analysis and data collection drives action, it is not necessarily high value-add. Actions need to include names and dates. Many projects are improved by using a CEDAC (cause and effect diagram with the addition of cards) method. While the cause and effect diagram (fishbone) tool is effective in understanding root cause, it is the additional use of cards that makes it so useful in the "improve" step within Six Sigma. The cards are used to keep track of results from specific changes made to the process. This requires control in limiting concurrent changes to best understand the direct relationship between cause and effect.

■ **Control** — The final step is the focus on sustainability and maintenance of the process improvements. If the process is improved through the first four steps as intended, normally the last step is to audit, document, and develop maintenance practices to ensure sustainability. This step follows the old adage that the work is not done until the paperwork is completed. Paperwork is often a nonvalue-add process component, but in the case of procedures, if these rules are not written down, they are not enforceable. Control is built from accountability and enforcement. In this step, tools supporting "mistake proofing" and an understanding of the importance of differentiating the difference between mistakes and defects are helpful.

DEFECTS

Six Sigma is more than just the DMAIC process. In high-performance Six Sigma environments, there are robust project funnels and processes to assign resource and engage people in regular process improvement. It would be completely remiss to avoid the topic of defects when discussing Six Sigma project management. Six Sigma will be defined more fully in Chapter 19, but it must be included in any project management discussion that, within the Six Sigma process, the objective is to limit process variation and allow predictable process results to the level of no more than 3.2 defects in a million opportunities. Project management within the Six Sigma structure requires complete and specific definitions of limits. Without these limits, it is impossible to determine the project effectiveness or measure sigma.

PROJECT MANAGEMENT STRUCTURE

The project funnel represents the potential projects and holes that need plugging to accomplish the business imperatives. The funnel requires a simple database system to collect the ideas and to approve or reject the projects and prioritize them for resource assignment. In high-performance Six Sigma environments, a "Six Sigma council" is normally established as a ruling board that oversees project prioritization and resource assignment or approval. This steering committee is often populated with the top-management staff of the organization or facility, allowing the highest level of support and commitment (Figure 13.9). A monthly review in a typical Six Sigma council meeting might include:

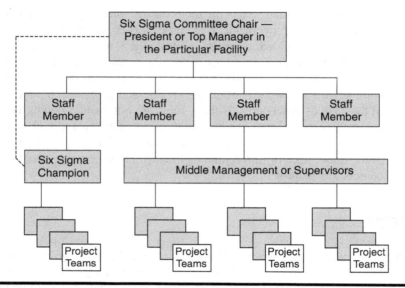

Figure 13.9.   Six Sigma Project Management Organization Structure.

- ■ Review of overall project impact in currency value
- ■ Review of the Six Sigma metrics — number of projects launched, number of projects completed, cycle time of projects
- ■ Review of top-priority projects chosen ahead of time for presentations
- ■ Review of any new certifications for black belts, etc.

Project management is a logical component of any improvement process and needs to be taken seriously. Doing so sends a major signal to the organization on the change component of normal work. Improvement becomes a condition of employment, not something we do when we get time.

> **Improvement becomes a condition of employment, not something we do when we get time.**

When project management gets the correct priority, it is easy to move a tremendous amount of change in an organization. Change is best determined and implemented by people who will have to own it after the implementation. Having the entire organization trained in project management skills and in good problem-solving techniques allows for efficient, timely, and effective change.

Because project management infrastructure is also a component of Class A, it makes sense to start with this type of structure even before the Six Sigma

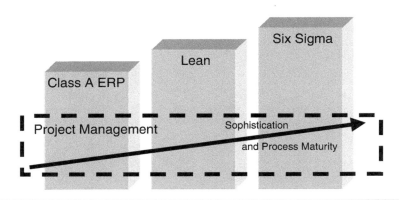

Figure 13.10. Project Management Integrations Through Class A, Lean, and Six Sigma.

maturity takes hold. The structure is the same except that the Class A champion is substituted for the Six Sigma champion.

Again, the idea here that is most valuable is not what you call project management, but that you are doing it. Project management is a common thread throughout *any* improvement activity or focus (Figure 13.10).

The project review certification requirement in Class A project management is one of the components of the Class A management system. There are many others. In Chapter 14, we will review the management system, part of which is directly attached to project management.

14

CLASS A ERP MANAGEMENT SYSTEM

The term "management system" has been used throughout this book and the weekly performance review has been referred to numerous times. Getting into the specifics of the Class A ERP management system is way past due. Do not be fooled by the sequence of topics in this book. The management system is the reason Class A works. Without it there would only be a set of metrics, which in itself does not drive change. High-performance companies, in every case, with no exceptions (how many times can you say something like that?), have robust management systems.

The management system in Class A has several elements. The *monthly* events are listed in Figure 14.1. In the monthly management system, the focus is on overall accountability and follow-up. It is in this space that management defines the priorities and makes sure these objectives are accomplished. The sales and operations planning (S&OP) process (covered in detail in Chapter 6) is a major element of policy decision making and risk management. Project management (covered in depth in Chapter 13) is also a non-negotiable component of high performance and, accordingly, Class A ERP certification. That would leave little to add here if it were not for the need to tie all of this together. It is communicating the *idea* of a management system that really gets people in the organization to understand what management is trying to accomplish by it.

The monthly management system is about setting policy and making decisions about the direction and tactics month to month. Top management owns the monthly management system space.

219

Figure 14.1.   Monthly Management System Elements in Class A ERP S&OP Project Management Review.

The monthly management system is about setting policy and making decisions about the direction and tactics month to month. Top management owns the monthly management system space.

The weekly management system is a bit more detailed and incorporates most of the rest of the organization. That would include all process owners, middle and line managers, and supervisors. The weekly management system requirements for Class A performance are weekly performance review, clear-to-build, and weekly project review.

WEEKLY PERFORMANCE REVIEW

The weekly performance review is one of the most powerful elements of the management system. With the exception of the S&OP process, there is no process more influential within the Class A ERP management system process. Done correctly, within a short time this meeting makes management into Class A ERP converts, and in every case I have been involved with, there is nothing that could stop the weekly review from happening. One of the main Class A ERP rules followed for the weekly performance review is that this meeting is never preempted. The weekly performance review has top priority for time management.

The weekly performance review meeting is best held on Tuesday afternoon to review the performance of the preceding week. As each process owner is to report the performance and actions driven from this process to the rest of the operations management staff, Tuesday has proven to be the best day; Monday is too early in the week to be fully prepared with analysis of last week's performance and misses. Tuesday afternoon is also seemingly the last point in the week where the week is still young but there is adequate time to prepare. Class A criteria require that this meeting be held consistently each week and do not dictate Tuesdays. As long as the meeting is at the same time each week, criteria are met. This eliminates the possibility of issues such as "I did not get the e-mail" or "I didn't know when the meeting was."

At a minimum, metrics normally reported in this format include, but are not limited to:

- Schedule adherence
- Schedule stability
- First-time quality
- Bill of material accuracy
- Inventory location record accuracy
- Item master accuracy (this can be more than one metric)
- Procurement process accuracy
- Shop floor accuracy
- Customer service

The weekly performance review process is an integral part of a robust management system that allows management to be involved in the follow-up of process control on a regular and predictable schedule. It helps to establish accountability for process ownership. Without it, process ownership has little or no real meaning to the organization and will not be effective. In a high-performance organization, as metrics are added to the deck, these metrics are also added to the weekly performance review. It becomes "how you run the business" in terms of process ownership and follow-up.

By using this type of forum for reporting progress and performance, the organization becomes consistent and effective at problem solving and continuous improvement. It should be used consistently for all performance metrics. As metrics are included in data analysis, it usually makes sense to have these metrics reported at the weekly performance review meeting.

The Agenda for the Weekly Performance Review Meeting

The weekly performance review meeting should be predictable and repeatable. This means that the meeting should be held the same time every week unless a holiday falls on that particular day. The agenda should be simple. Each process owner reports his or her progress for last week. Progress includes all elements on the quad chart explained in the next few pages (see also Figure 14.2). Questions are appropriate from the other team members if the process owner is not reporting performance to acceptable levels. The objective is not to "pile on" problem process owners, but instead to offer help. Help can be in the form of resources, ideas, a change in handoffs, etc.

- **Top left of quad chart** — In this example, the performance of inventory accuracy is being reported. This same format would be used by all process owners with just a title change. In the top left-hand quadrant, the daily performance is reported along with the weekly overall performance summary. Most Class A metrics are expressed in percentage points. Performance number is to be reported as a daily and as a weekly total.
- **Top right of quad chart** — The trend chart is displayed in the top right-hand quadrant (Figure 14.2). This is important to determine if the

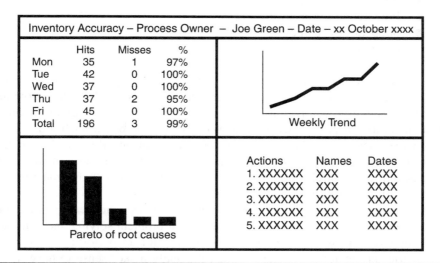

Figure 14.2. Quad Chart Performance Reporting Format.

process is affecting the performance positively. It is not clear if progress is being accomplished successfully if only the performance for the current period is reported. The trend chart should cover a minimum of five to six weeks of performance data. The trend graph reported in this quadrant is typically weekly data points only. This creates a smoothing effect that can give a better and truer trend summary.

■ **Bottom left of quad chart** — A Pareto chart showing the reasons for misses will allow the organization to focus energy and resource on the most important barriers to successful accuracy. This is especially important early in process focus and measurement when process is affected by the most sources. The best-performing organizations focus resource on the worst root causes, allowing the biggest return for their investment. It is in this quadrant where the biggest challenges will come as the culture migrates. Process owners often will struggle with the definition of root cause. Management normally has to combat "categories" versus real root cause. Without root cause, it is impossible to really eliminate the variation causing performance below goals.

■ **Bottom right of quad chart** — Process improvement comes only by driving change through actions. By institutionalizing the action process and reporting it, follow-up is much more systematic and predictable. The root cause bar at the left side of the Pareto chart should have actions easily linked on the right-hand side of the quad chart (Figure 14.2). In the beginning of the weekly performance implementation, it is common that root cause will not be well understood. The old rule of asking "why" five times helps get to root cause. Another simple root cause rule is to ask why until the answer is actionable.

> In the beginning of the weekly performance implementation, it is common that root cause will not be well understood.

During the weekly performance review meeting, process owners need to stay focused on root cause to avoid excuses. This will drive proper corrective actions and will reduce the cost needed to eliminate the barriers to high performance. Management has to play a role in this if it is to become a required behavior and a predictable component of the management system. In most businesses, the root causes today are the same ones that existed two weeks ago and are the same ones that existed two years ago. These root causes must be eliminated.

> In most businesses, the root causes today are the same ones that existed two weeks ago and are the same ones that existed two years ago.

CLEAR-TO-BUILD

The clear-to-build (CTB) process was reviewed in Chapter 7, but because it is an important management system event and because it is a requirement of Class A ERP performance and certification, a summary is appropriately included in this chapter.

The CTB process consists mainly of a commitment from production and procurement managers to the detailed master production schedule (MPS) for the coming week. The master scheduler distributes the new MPS late in the week. In most businesses, this is on either Thursday evening or Friday morning. On Friday, the key players are expected to either confirm their commitment and confidence at a gathering or, in more mature, experienced, and well-master-scheduled environments, no meeting is held. Instead, the process owners of the plan execution only report back if they require adjustments prior to commitment.

In one multiplant bakery I worked with, the master scheduler in Baltimore used to send a schedule to each plant manager at the other locations on Friday morning. If no response was received by 2 P.M. on Friday, commitment was assumed.

WEEKLY PROJECT REVIEW

Class A criteria do not specifically dictate the interval for project review. In many organizations, high-level management reviews most projects at least once a week. It totally depends on how many projects there are in process, how important the outcomes are to management (high-profile projects are reviewed more frequently), or what the impact or risks are.

DAILY MANAGEMENT SYSTEM EVENTS

It has been said many times that if you get the day right, the week will be fine. If the week goes well, you do not have to think about the month. This is probably a little simplified in terms of the real world, but the spirit of this statement is correct. Since some businesses still have an "end-of-the-month crunch," there is probably something to be learned from daily management systems. In Class A ERP, there are a few that are included in the criteria: daily schedule alignment, daily walk-around, visible factory boards, and shift change communication.

Daily Schedule Alignment

In Class A ERP, it is expected that all information in the system will be as current as there is knowledge in the business. Every time a situation changes, the system should reflect that change. That means there will be almost no past-due requirements in the ERP business system at any one time. There is plenty of process variation in the world; in fact, as all Six Sigma experts know, *all* processes have variation. This being the case, and given the importance of keeping the system up to date, alignment needs to happen often, at least every day.

The alignment is as simple as an assessment of production schedule adherence from the previous day and adjusting accordingly. In some businesses, there is no extra capacity. This is often true in high-volume, repetitive manufacturing. When the schedule is missed, the time and associated capacity are lost. The schedule needs to reflect that change. By the same token, if the business has some flexibility, misses from the prior day can sometimes be made up by working some overtime, doubling up on resource, etc. In any case, the schedule should reflect the latest expectations. When the MPS is updated, material requirements planning can align signals to the supply chain accordingly, and both inventory and shortages are minimized.

Class A criteria require this alignment in the system to happen at least once a week, but it is done every day in most high-performance businesses. When schedules are kept accurate and realistic, the result is minimal system noise. The schedule should be reviewed daily by the production managers, materials management, and master scheduling. This often happens in a daily communication meeting first thing each morning, sometimes through phone calls or a short update meeting. During this meeting, the focus is on alignment — how many units were completed yesterday on the line and how many in the sequence we expect to be able to finish today. This is discussed on each line, agreement is reached, and the master schedule is updated accordingly. This process is a "mini-CTB" process. In high-performance organizations, this can go quite fast. If there are a lot of problems, however, this process can get slowed down. In many manufacturing plants, daily schedule white boards are bolted to the machine lines. In the daily schedule alignment process, the boards (visible factory schedule boards) are updated each day with yesterday's performance and any changes to the schedule. The boards should be big enough to be read easily when walking by. The master scheduler collects and confirms any changes, and the new schedule is updated into the system.

Variation can come in both directions. Poor schedules or poor execution that is difficult to schedule accurately can result in schedule variation to both the high and low sides. Also keep in mind that in a Class A manufacturing organization, this may only involve 5 percent of the business on any one day. The rest of the lines or production should be running at 100 percent of schedule.

Daily Walk-Around

The daily walk-around is exactly what it sounds like. Management reviews schedule adherence each day by walking around the facility to each line supervisor to get an understanding of the progress made the day prior and what the expectations are for the current day. This visibility helps in several ways:

- It helps to enforce the belief that schedules need to be accurate.
- It keeps management in the loop in terms of any variation development in manufacturing at the line level.
- It allows management to acknowledge good problem-solving behavior.
- It gets management in the loop so it can help with more difficult issues when appropriate.

One example always come to mind when I talk about the daily walk-around. While working in China, I was doing the final Class A audit in a plant. In this plant, the daily walk-around consisted of four people: the plant manager, the production manager, the master scheduler, and the engineering manager. These four managers would meet in the factory each morning at the same time to tour the facility and talk to each line supervisor. In this particular facility, there were no middle managers. All of the line supervisors reported to the production manager. Line supervisors were team leaders who actually worked on the line as team members most of the time.

The reports I observed that morning during the audit were interesting. I met the others in the designated aisle at 8 A.M. We walked to the first work cell. The supervisor stopped his work and greeted us. The board showed no misses from the previous day, and the supervisor informed us that he did not expect to miss any scheduled product on the current day. The same thing was reported on the second line, but when we reached the third line, I started to perk up. Line three had missed several units the night before. Its performance was posted at 92 percent, which is unacceptable in a Class A organization. As reported, the following sequence of events happened.

The third shift had trouble with a machining center. A motor had burned up on the machine, thus stopping production. The late-night shift supervisor went into maintenance, got a new motor from maintenance stock, and assembled it to the machining center himself. The machine was restarted and production was able to minimize loss to the schedule. After production was back on track, the supervisor went back into maintenance to *fix* the problem. He pulled the records for the machine and found that the scheduled preventative maintenance (PM) was to be done in about two weeks. This PM would have included taking the motor apart, checking the motor brushes, and blowing out any carbon dust. All of this would have probably prevented the unscheduled machine-down situation experienced that night. On that same shift, the PM schedule was changed to a shorter interval by the supervisor, with communication left for the maintenance manager.

I do not see many supervisors who are that well trained and/or empowered to repair their own equipment and who also understand that this was not the real *fix* to the problem. This operator had been trained to ask "How do I stop this from ever happening again?" This is clearly the spirit of Class A performance. The walk-around is designed to keep management in the loop and to keep accountability at the line level. Given safety and training concerns, it is not expected that every team would be able to do its own machine repairs, but understanding root cause can be a big help to any and every process owner.

The daily walk-around is helpful. Several processes can be reviewed during this tour, such as schedule adherence, inventory accuracy, and housekeeping, and management can also observe progress on recent projects. With the busy schedules of most plant managers, this is a favorite time for many. It forces them to spend valuable time on the factory floor, and when things are running as they should, it also gives them the opportunity to show appreciation by thanking people and acknowledging good performance in public.

Visible Factory Boards

To keep the schedule well communicated and understood, most high-performance factories are not relying on paper schedules hung by the machines or work cells. Instead, these well-managed companies are using large white boards with the daily schedule information posted for all to easily read (see Figure 14.3).

Variations are acceptable to the visible factory board layout. Many organizations also put safety and first-time quality performance on the board, and this also has been helpful. Some organizations choose to do schedule checks more than once a day. One client I am currently working with chooses to have schedules checked twice a shift; on the day shift, those schedule checks happen

Schedule ___/___/___			Schedule ___/___/___			Perf
Product	Quantity		Product	Quantity		MTD
	Plan	Act		Plan	Act	
XX	23		XX	78		
XX	111		XX	99		
XX	5		XX	455		
XX	589		XX	4		
XX	45		XX	135		
Total	773			771		
Performance XX%			Performance XX%			
Open Issues			Open Issues			

Figure 14.3. Visible Factory Board.

at 10 A.M. and 2 P.M. The benefit of this management system is the communication value to ensure everyone is on the same page when they start the day. It also helps to underline the importance of making schedules as planned.

Shift Change Communication

Each time a shift change happens, there is an opportunity to lose continuity in production. Each shift means new people come in to replace the people who are presently working on the schedule. In some businesses, it means that the processes and speeds will vary. In high-performance businesses, this is not the case. Much effort is devoted to keeping operators educated and trained to understand what the knobs on the machine really do and to setting up standard procedures to be followed by all operators. In one client organization, manufacturing measures all changes to process. Every time there is a change from the standard process settings required to bring product into spec, a miss occurs on the line's metric. The metric is reported at the weekly performance review meeting and receives a lot of focus. This helps to focus operator awareness on root cause. Many things have been learned from this. I highly recommend it.

Shift communication is difficult. People on the leaving shift are tired and ready to go home. Hanging around to answer questions is not always the most effective means of communication. Instead, many high-performance companies will have formal communication processes established. Formal and visible communication boards on the line are helpful. Having a diary-type log in the cell is also helpful. Class A criteria suggest that the process be designed for your environment, but it should be consistent throughout the factory. This is just common sense and worth the extra time to establish good habits.

OTHER MANAGEMENT SYSTEMS IN THE BUSINESS

Management systems are everywhere in high-performance businesses. In businesses where these elements are second nature and woven into the fiber of the organization, there is no discussion about management systems. They just happen. Whenever there is a process change, the documentation is updated automatically because it just makes sense. If a metric is changed or a new metric is born out of need, the natural first thing to do is to include it in the weekly performance review system. All projects are naturally included in the project review. In these companies, there is just no decision — it is obvious. This describes the Class A ERP thinking and environment. Predictable process only comes from a combination of two things: a good plan and good execution. One without the other spells failure at one level or another. There are no exceptions. Thought and planning on management system elements can foster a culture where Class A ERP thinking happens naturally. I have met many people years after we first worked together who are still following the basics of Class A thinking, even if their current mission is strictly in the Six Sigma space. It is just good process.

In the next chapter, the biggest effect on culture change will be discussed: education and training. The only way I know to change culture quickly is to make sure everyone is on the same desired wavelength as soon as possible. Education and training, done appropriately, are the road to that end.

15

EDUCATION AND TRAINING

People confuse education with training. These are two separate areas, and the deliverables, while both valuable, are not equal. Combined, these two investments can help change a culture in an organization. Education is about getting employees to understand why change is important, and training teaches them how to do the needed tasks.

CLASS A ERP EDUCATION

Class A ERP education starts on day one of the implementation. It starts with a communication session that describes the "should be" position and compares it to the "is" condition. This description should include performance goals and some benchmarking to ensure that the objectives are achievable. Even before the implementation process begins, however, education has to happen at the highest levels in the organization. This can be some of the most important education (see Figure 15.1).

Top-Management Education

The first education that needs to happen in an organization ready for Class A implementation is with the top management. Without management's understanding and support, the process will be very difficult to accomplish, no matter

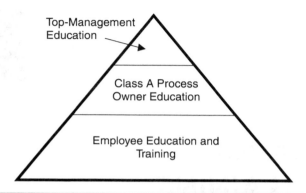

Figure 15.1. Education Pyramid for Class A ERP Implementation.

how much commitment there is from the ranks. The truth is, it is very easy to fail Class A, lean, or Six Sigma implementations without management support. It happens every day throughout the world. Employees always take their cues from the top-management team, especially the CEO or president.

The education of top management has to be done on their terms, when they are ready, and at a convenient time. It is best to get it done in one short burst, as it is difficult to get these busy people in a room together without interruptions. Normally, the initial Class A ERP introduction can be delivered in about six to eight hours depending on the exercises incorporated into the session. The deliverable should include a common glossary of terms, a methodology for implementation, and an understanding of roles and expectations for each of the levels within the organization. If the Class A project manager has been chosen prior to this initial education, it is good to have this person attend the first education along with top management.

> If the Class A project manager has been chosen prior to this initial education, it is good to have this person attend the first education along with top management.

This education would normally cover topics of sales and operations planning, operations planning, execution of schedules, and data accuracy. It would also briefly cover expectations from the certification process following successful implementation. Credibility is key in the delivery, so it is important to choose the delivery agent carefully and make sure they have successes to draw from in their examples. Top-management education is almost always delivered by outside expert resources.

Educating the Process Owners

The core of the Class A implementation team is made up of the process owners of the core processes such as sales, engineering, master scheduling, production, materials, purchasing, inventory control, and shipping. These middle managers are intimately involved in the details of selling product, converting raw material into saleable product, and shipping product to customers. If these people can become one voice for change and begin to approach the implementation of discipline and management systems from the same perspective, the process moves more efficiently and becomes effective much more quickly. This is how culture shifts take place in organizations. Cultural change takes place when the opinions shared by many rally around new beliefs. When understood thoroughly, Class A ERP is not very controversial to most management people. It is the understanding that makes it easily accepted. Nothing can help this shared understanding any more quickly than education.

The right way to do the education at the middle-management level is to have it delivered by someone who has been at the heart of one or more successful ERP implementations and has seen it work well. The detail has to be concise enough to satisfy skeptical minds. Because of these requirements, the initial education is often done by an outside expert. This would include detailed overviews of all of the processes within the Class A business model, from top-management planning to plan execution and data accuracy.

The education of the core team does not stop at this initial introduction. Many teams are involved with education beyond the initial delivery. If the topics are not kept in front of the team members frequently, much is forgotten from the initial introduction. Follow-up exercises can be very helpful, as are instructional tools from the American Production and Inventory Control Society (APICS), APICS certification goals, and video tools that are available on the market.

Education of the Masses

Ultimately, Class A ERP will impact everyone in the company, especially if it is implemented correctly. This involvement will come through helping to determine root cause, collecting data, paying more attention to details concerning process discipline, working on process improvement projects, reviewing visible factory boards, and even reporting to the weekly performance review meeting. For all the same reasons that top and middle management need to be well versed in the cultural shift about to happen, the factory and office workers need to be clued in as well. This happens through exposure to the ideas during Class A ERP education and training.

Educating the employee base too soon, however, can create expectations and disappointment if people do not see management's actions following shortly thereafter.

Educating the employee base too soon, however, can create expectations and disappointment if people do not see management's actions following shortly thereafter.

It is good to get the management team well down the path and starting to implement actions prior to the employee base education. Once management starts to take actions, such as implementing metrics, initiating the daily walk-around, etc., education must be delivered plant- and company-wide. One of the best ways to deliver this education is with the core group from the implementation team and/or the line managers. When these managers have to deliver education to their departments, the message is strong and succinct. Again, video education in the areas of ERP can be useful, as can APICS materials. Outside resources can be used, but sometimes are overkill at the lowest levels in the organization. Exceptions to this are specialized areas such as master scheduling, inventory control, and warehouse management. In these areas, Class A standards can be delivered effectively using outside resources. Each case needs to be evaluated for budget, delivery skills on-site, and employee needs and appetite.

TRAINING REQUIREMENTS

Training cannot begin until the processes are reviewed and either approved or updated to reflect Class A process requirements. This is especially true in inventory transaction process training, performance measurement training, and other processes affected greatly by Class A influences. Training is teaching people "how" to accomplish tasks and processes successfully. The sequence is:

1. Assess the "is" condition
2. Determine the "should be" condition using proper problem-solving techniques
3. Implement changes required in the process
4. Document the changes in procedures
5. Deliver the training to ensure sustainability of process

If you look at these steps in the context of the Six Sigma DMAIC (Define, Measure, Analyze, Improve, Control) process, you will notice the same logic used to determine the required steps. Documentation is a necessary evil and is part of the process control requirements of Class A ERP, International Organization for Standardization (ISO), and Six Sigma. Unfortunately, if it is not written down, it is probably impossible to sustain or audit compliance.

Documentation as a Training Tool

No one likes the notion of adding paperwork to any operation or transaction, but as repeatability of process becomes the target, documentation becomes necessary. Without written agreement, humans tend to remember things differently. This is historically true in everything from land boundary disputes to factory transaction procedures.

If two people in a room agree on a process rule, each leaves the room with an idea of what the agreement was. Each of the players has an experience filter through which he or she listens to process discussions. Because of these filters, the interpretation of the agreement can and often will vary. Therein lies the need for documentation. Without documentation, procedures can deviate and evolve in a direction unrelated to the original objectives of the process. The purpose of documentation is to perform or support the following value-add advantages:

- **To provide consistency in the business** — The documentation of processes by itself does little to increase consistency. The combined efforts of auditing and insistence that documentation be correct are the vehicles that instill consistency in the manufacturing cycle. The objective should be repeatability of a process with predictability of performance. Many companies pursue ISO certification for exactly that reason. The ISO standard is a globally accepted manufacturing consistency standard originally made widely known by business organizations in Europe and suppliers to European companies. ISO structure and discipline can provide the necessary infrastructure to control and audit documentation and procedures for consistency. In an ISO environment, the data accuracy procedures developed from the inventory record accuracy initiative can fit appropriately into the ISO process of controlling documentation, although ISO certification is certainly not a prerequisite to accuracy. This internationally recognized standard has been adopted by many companies worldwide. Done properly, many such companies find at least the standardized ISO format for procedures helpful in their quest for data integrity. *Done properly* is the key element!

- **To gain control of the processes** — In most businesses, changing the culture of an organization is a significant and challenging undertaking. When people hold certain ideals as important in their lives, these ideals begin to impact their surroundings. This relates to certain expectations in consistency and predictability. For example, in some businesses, expediting that includes shortcuts and behaviors that skip steps can be viewed as good behavior, even rewarded in some cases. In many of these organizations, a good understanding of the process and procedure is inadequate, resulting in a lack of concern for the steps in the procedure.

■ **Process mapping** — Documentation plays a part in this transition. If a department or division manager has no real detailed understanding of a specific process, he or she is at the mercy of the people who hold the knowledge base. While this may seem reasonable to some, it can put a business in a very precarious position. Process mapping, the methodology of documenting the steps and flow of a process using symbols, is an effective tool for pictorially representing the "is" condition of the current process as well as visually describing the desired procedure. Process maps can add great value. Gaining control over a business requires an understanding of the processes as they exist today. Documentation allows for this understanding and creates the baseline for improvement. In organizations where management understands the benefits of documentation, process maps always provide great value.

■ **Training guides** — One major responsibility of standard operating procedures (SOPs) is to verbalize or document the process maps. Once the process is documented correctly, showing all of the decision points in flow chart form, it can easily be turned into a training document. This is done by simply writing down the actions and decisions made during the process as described in the process map. The result is an SOP that describes not only why but how certain functions are performed. There is, of course, more to the SOP than just the verbalization of the process map. Additionally, there are other necessary information or communication items that are helpful in achieving global understanding and acknowledgment. These communication items usually include boundaries of scope, expected outcome or objective of the procedure (sometimes referred to as "deliverables"), tools required, performance metrics associated with the procedure, a listing of other related SOPs, samples of documents used in the process, author(s), authorization(s), and date last reviewed. This all results in perfect training materials.

■ **To enable continuous improvement** — Of the documentation purposes listed here, enabling continuous improvement is probably the most important. Once the organization becomes process oriented, opportunity is evaluated in terms of waste elimination or process variation reduction as it applies to process and procedure. Each step in the current "is" condition can be evaluated to determine if it adds value to the process. Continuous improvement results from simplification efforts to move the process toward the "should be" condition. The expectation should include a procedural review at reasonable intervals. This normally means every document must be reviewed every six months for accuracy. If this is done regularly, most organizations find that they can review just a few each day and still stay current with the requirement. If each process owner reviews his or her own documentation, it is a fairly efficient process. The review requirement can be as simple as reviewing a few procedures each month for accuracy.

Accessibility to Documentation

Obviously, computers and system networking are having a major impact on business processes today. These tools are helping to take time out of processes and are aiding the reduction of customer service cycle times through efficiency gains. In many cases, accuracy is impacted positively. Because of this exciting positive effect, more and more applications for computers are being developed daily. Most businesses today are networked through connected hardware and software tools. This allows the opportunity to have SOPs accessible to all who are on the network. While this may seem adequate, maybe even advantageous, be reminded that the limitation of this practice is accessibility to the system. People on the factory floor or in the warehouse need access to SOPs just as frequently as do the office personnel. If the culture is to know the procedure and follow it, access must be easy for all. Many companies have a complete hard copy set of all procedures available in a central location (such as the lunchroom) for people who might not have access to the computer network system. If the cafeteria does not lend itself as a good reference library location, other special documentation areas can be established in the factory or warehouse. This centralized and well-advertised location is also a good place to have company policy documents accessible, such as human resource practices and policies affecting employees.

If documents are available in hard copy form as described, a procedure also must be instituted to provide for the timely placement of procedures when they are revised. The format and procedural review typically associated with an ISO certification can provide a guideline for review and revision.

Standard Operating Procedures

It is recommended that as documentation is initiated in an organization, a standard format be adopted. If your organization has done work in ISO requirements, you are already familiar with this requirement. Having a template supports consistency and makes sure that all the desired elements are resident in each procedural document. The following list of topics is reasonably complete and these topics are normally found in a well-written SOP template.

Define the Boundaries or Scope

The art of process documentation includes sensitivity for linkages to other aspects of the process or business. Each process and corresponding SOP has a reason for existence. The SOP should refer to other SOPs that address linkages and dependencies. These linkages identify waste opportunities.

By the same token, it should be noted in the SOP that connecting processes are linked with the output of the process described. For every process, there should be a driver/cause and a result/effect. Again, it is this cause and effect relationship understanding that will enable a team of educated employees to convert opportunities into business improvement. Many companies connect these processes visually. Using string, they attach to the wall of a conference room each process as it was documented. They then connect the effects to causes. When completed, all four walls are usually decorated with the complex process of order fulfillment in the facility. It also becomes obvious where the holes in the process are and where linkages to existing processes reside. Visualizing it may help to appreciate mapping as a powerful tool. Once this exercise has been completed, the definitions of scope or boundaries of the SOP are more easily stated. By defining these boundaries, it becomes easier to train and audit the procedure later.

The wording to accomplish the scope might simply state something like: "The procedure starts with the generation of a pick requirement in the MRP system and ends with the actual movement of material from the stockroom to production." A list of related SOPs would be referenced for process linkage.

Describe Why This Procedure Exists

Describing why the procedure exists can be accomplished in the objective statement and does not necessarily have to be a separate paragraph of the SOP. If this definition is missing, too many times the procedure can be misunderstood and manipulated for totally different purposes. The explanation of why the process exists is often an opportunity for further improvement.

Describe the Objective or Deliverable

Describing why a procedure exists (above) and describing what the deliverable is are two different things. Both help to lock the spirit of the procedure for the future in case of interpretation. The deliverable should be measurable. In an intracompany environment, an example of a bill of lading SOP for internal transports (plant to plant) might be "to provide proper documentation for all shipments between divisions in different geographic locations, to provide accurate and timely visibility to material control analysts and accounting personnel of intracompany material movement." By listing the objective, the author is tested as to whether the objective is worthwhile and is given expectations on how to verify if the procedure is working properly.

List Steps in the Process

List the process steps in order. Capture all decision points and alternatives that are in the procedure in real life. It is these alternatives and decision points that give the SOP its real instructional value. This should quickly remind you of the previous discussion on process mapping. In this section, many organizations actually attach a process map to best fulfill this documentation requirement.

How many times have you wanted to know something about a software application and found the help tool to be of no help at all because the particular feature or option had been left out of the instructions? Have several team members review the list for accuracy upon completion of the draft.

List Special Tools Required

This list of "special tools" would normally include items required for the successful completion of the procedure. In a warehouse, these might include items such as a computer terminal, software tools, fixtures, chain sling, forklift truck, scales, special measuring equipment, gauges, etc. This also helps to eliminate nonrequired instruments/tools from the specific area.

It can be interesting to audit the tool section of procedures when doing 5-S housekeeping and workplace organization exercises. If it is not in the documentation, ask if it is needed. If it is, update the SOP; if not, get rid of it!

Reference Performance Measurements

Most processes should have *at least* one performance measurement ensuring process control. Work instructions and/or material conversion processes also might have quality standards as performance measurements and/or efficiency measurements.

The SOP should include a definition of that measurement process and, if available, a cross-reference to the SOP that defines the measurement in detail if it is not described in the procedure itself. If applicable, an example of the measurement calculation also should be attached. This is a very important element of procedures that is often left out. The measurement makes auditing very easy and can lead to early detection of noncompliance.

Reference Other Standard Operating Procedures or Work Instructions That Pertain to This Process

If there are other SOPs or pertinent documents that have linkages to the current SOP, they should be referenced by a document control number. This is typically quite common, especially if the processes have defined inputs and outputs.

Attach Sample Documents and/or Sample Computer Screens

Many processes have control documents and/or computer screens that are used in the procedure. Make sure both are referenced in the procedure and have examples attached in the form of pictures or graphics. This requires a serial number or control number to be assigned to each document. This serial number would be found on both the procedure and the attached referenced document. When these are logged into a database, referencing becomes an especially powerful tool. For example, when changes or updates are being made to a procedure, having full reference to related documents helps to ensure complete process revision integrity.

The Author

Knowing who wrote a specific procedure can be extremely helpful when clarification is required. If more than one person was involved, the one responsible for the process itself should be referenced. Avoid multiple names in this section. Process ownership is important for proper accountability. The author is often the process owner or a delegate of the process owner.

Who Authorized It (With a Signature Field)

The authorizing signature is important for controlling the variability in the overall process. SOPs should not be changed at random or by people who do not understand the impact on other parts of the organization or overall process. The authorizing signature should be someone with direct knowledge and authority over the area addressed by the procedure. This would normally be the inventory manager or materials manager in inventory control procedures.

In some organizations, the plant manager signs all procedures as a sign of support for required disciplines. While the spirit of this decision is valid, the realities of the effectiveness can be questionable. Having the plant manager sign numerous authorizations without really knowing the process in question is not a value-add activity and should be avoided. Many times in these instances, the plant manager does not even read them; he or she takes the word of the author. This does not necessarily contribute to a value-add process.

Date Written

The date that the SOP was written should be a required field for reference purposes. This can help with investigations and audit trails. This date does not change even as the SOP is updated and revised.

Date Last Audited

Procedures should be audited on a regular basis. This means that just as a company might audit the financial records, inventory record accuracy, or manufacturing bill of material accuracy, procedures also should be audited. The date in this field should reflect the last time the procedure was verified as accurate. It does not mean that there is an expectation to change every procedure every six months. It does mean procedures are audited for accuracy on a timely basis.

What many companies do is post a verification audit for every procedure every six months. The process owner is responsible for this audit, and if no changes are required, he or she simply signifies process accuracy and the audited date is revised. If this is done regularly, most process owners will have only a few each week to look at and it does not become insurmountable to keep up with. If, on the other hand, it is ignored until the end of the year or there is an expectation that all SOPs will be reviewed in a short time frame, it can become a nightmare to achieve.

Documentation Control

Once documentation is created and processes are defined properly and are being executed, revision control must also become a priority. Similar to the requirements of engineering change control, management must provide tight accuracy and effectivity controls on SOPs. This control is necessary to protect the investment that has been made in understanding, improving, and documenting the process maps and procedures of the organization's activities.

Documentation control should be kept simple. Process owners, the people who are responsible for the particular process described by the document, are the "managers or controllers" of change within that process. This puts the process owner in the position of auditing, updating, authorizing, and communicating process details and controls defined by an SOP. Document control becomes part of the oversight requirements of the process owners. A document control procedure would typically include the process to deal with the submission of ideas pertaining to improvements, investigation, and/or testing of new process designs, procedure authorization, revision communication, training when required, and finally filing of the revised method. The document control process itself should have a written procedure to limit variability in this important process.

All individuals directly involved in the procedures should be trained in document control procedures just as they are trained in their respective processes. Each procedure and map should have a control number assigned. This control number, along with the initiation, revision, or audit date, acts as the revision-sequencing tool. As they are replaced by newer procedure revisions, previous procedures should be discarded to eliminate confusion.

APICS in the Workplace and Other Outside Resources

There are materials available in the marketplace that can support the training requirements in any business, regardless of size and budget. APICS is the professional society that is designated as the keeper of the body of knowledge for many of the processes being discussed in this book and many others like it. Networking with other companies like yours can be helpful, and groups like APICS, purchasing manager societies, and manufacturing engineering societies can provide great value if you are willing to put the time and effort into them.

Certification goals such as those offered by APICS through its CPIM (Certified in Production and Inventory Management) program are very beneficial, not just to employee resumes, but to the company as people increase their learning. APICS certification is especially useful as many employees will willingly work on this on their own time and classes are often conducted off-hours. The only expense to the business is often the course fees and certification exam fees.

In organizations that use APICS wisely, personal CPIM certification becomes a culturally desirable label as a differentiator within the business. At The Raymond Corporation, APICS certification was supported to a level that influenced both the CFO and VP of operations to become APICS certified. This sent a strong signal throughout the organization. Before long, we had people from engineering, accounting, and even human resources studying and taking the exams. The business was much better for it.

DVD EDUCATION FOR ERP

There are other tools available for ERP education and training. These tools seem to get easier to use each year. One particular example I have used successfully a DVD product called "ERP and Supply Chain Management," available from Buker, Inc., a business management consulting firm. Tools like this one allow first-line managers to deliver education and training on the topics under the Class A umbrella with help from teachers' guides and workbooks. These tools are also designed to be resident on a central server and used by anyone in the organization as required.

The reason I have tended to use tools like this one in many implementations I have been involved with is because of the complexity of the topic and the wide-ranging experience levels managers have to deal with in an organization. With a modularized tool, the instructors can pick and choose whatever they find to be needed with their specific audience and skill requirement. While this may sound like an advertisement for the product, the truth is that I have used it successfully. It can be a powerful tool for organizations committed to the Class A ERP process. Every company has to assess its needs and match them with budget and appetite for material and content.

16

CLASS A ERP PERFORMANCE METRICS

Some readers will see the title of this chapter in the table of contents, at first glance, and will start to read the book here. In some ways, I can understand. Metrics are the drivers and source of needed information. High-performance companies do not run without them, and poorly managed and performing companies often have the wrong metrics in place or none at all. This makes performance measurements a necessary part of the feedback loop in businesses that are serious about improvement. In many ways, metrics are the keys to the palace. Class A ERP is built around metrics and the management system to keep the metrics changing and fresh.

BAROMETRIC AND DIAGNOSTIC MEASURES

There are two types of metrics in a Class A process (Figure 16.1). Many years ago at The Raymond Corporation, one of the first and best master schedulers I have worked with determined that the normal Class A metrics that are always appropriate in every business are "barometric" measures — they tell you if a storm front is coming. These metrics do not necessarily tell you what the cause is; they only point you in the right direction. The barometric measures should be maintained even when the performance level is sustained for many months. The reason these metrics continue to exist is for auto-audit purposes — understanding things are still in control from that perspective.

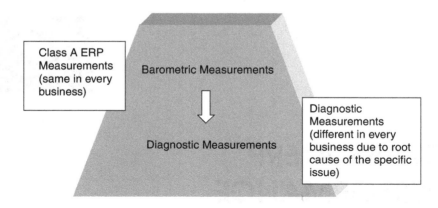

Figure 16.1. Barometric Measures Versus Diagnostic Measures.

There is another level of metrics required in every business called "diagnostic" measures. These measurements can change from business to business and should. These metrics are driven from root cause, seeking and focusing on areas of weakness in a specific situation. Environment, skills, process variation — these things are different in different companies and therefore will require different metrics. That is why they are referred to as diagnostic measures.

In Class A ERP, the measures correspond with the ERP business model plus the two topics of quality and safety. There is a barometric-type metric for each of the darkened boxes in Figure 16.2.

The list of metrics for Class A ERP is as follows:

1. **Business plan** — Accuracy of the monthly profit plan of the business by product family. If the facility is a cost center, the accuracy of the budget becomes the base for accuracy.
2. **Demand plan (forecast)** — Accuracy of the monthly demand forecast by product family.
3. **Operations plan** — Accuracy of the monthly capacity plan by product family.
4. **Master schedule** — Accuracy of the weekly detailed schedule by line.
5. **Material plan** — Percent of orders that are released with full lead time to suppliers. This is measured to the current lead-time field in the item master.
6. **Schedule stability** — Percent of orders that have completion dates revised within the fixed period time fence.
7. **Inventory location balance accuracy** — Percent of location balances that are perfect.
8. **Bill of material (BOM) accuracy** — Percent of BOMs that are perfect.

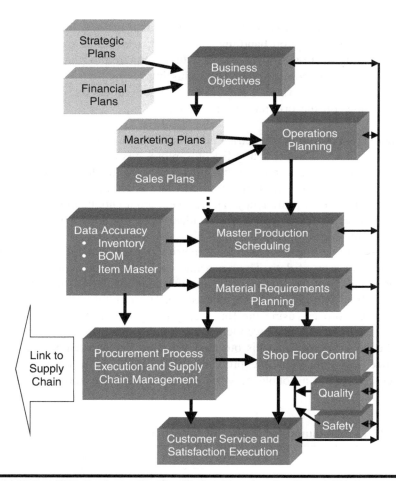

Figure 16.2. ERP Business System Model Showing Quality and Safety.

9. **Item master file accuracy** — Accuracy percentage of other item master fields that come up during the initial assessment of Class A ERP weaknesses. This can include routing records, lead-time records, cost standards, etc.

10. **Procurement process accuracy** — Percent of complete orders that are received from suppliers on the day (or hour) currently scheduled in the ERP business system.

11. **Shop floor control accuracy** — Percent of complete internal orders that are completed on the day (or hour) currently scheduled in the ERP business system.

12. **First-time quality** — Percent of units that make it through a process without exception handling, movement, rework, or rejection.
13. **Safety** — Lost-time accidents actual compared to plan. Minimum acceptable is company objective plus improving.
14. **Customer service** — Percent of complete orders that ship to the customer on the original promise date.

There are a multitude of metrics that would be found in a high-performance Class A business. The preceding list only represents the barometric measures. These will be the same in every business, from plastics to automobiles, from paint to capital equipment manufacturing. Class A requires these metrics to be at minimum levels of proficiency. These levels should not be misconstrued with high performance. Table 16.1 gives an idea of the minimums and what would be considered high performance.

The key to getting the value from metrics is really not the metric itself or even the collection of data. The real value comes from the action driven from the data. One client has four plants and a fifth being planned. The CEO is really intuitive about what needs to be focused on. He initially started a continuous improvement process using the four-step journey (see Figure 16.3).

His first comment when we met to discuss his organization's needs was that he believed in metrics and his business was full of measurements. Even performance was good, but one problem still existed — the performance had stopped improving a couple of years prior. Profits were good and customer

Table 16.1. Minimums for Metric Performance

Metric	Percent	
	Minimum Class A	High Performance
Business plan	95	100
Demand plan (forecast)	85	90
Operations plan	95	98
Master schedule	95	98
Material plan	95	95
Schedule stability	95	95
Inventory location balance accuracy	95	98
BOM accuracy	98	99
Item master file accuracy	95	98
Procurement process accuracy	95	98
Shop floor control accuracy	95	98
First-time quality	95	99
Safety	Plan plus improving	Zero accidents
Customer service	95	98

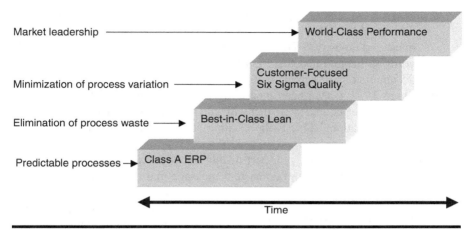

Figure 16.3. Journey to World-Class Performance.

service was great, but the numbers were not growing at a rate that satisfied his high standards. This is exactly the attitude that makes change easy and fun. Top management sees the need for continuous improvement and is never fully satisfied. Class A ERP, the first step in the journey, establishes the management systems necessary to keep progress moving. It creates the expectations of improvement day after day.

STARTING THE METRIC PROCESS IN CLASS A

Some businesses measure every process and this expected action has become part of the culture. In other businesses, as metrics are added to the deck, resistance is detected. Performance measurement in high-performance organizations is not "something they do"; it is "the way they think."

Typically, there is not a great desire to measure performance throughout the business in the beginning stages of implementation. As is human nature, people often see this change as a threat to their security or even a violation of their honor or trust and think, "Why do you want to measure my performance anyway?" or "Do you not trust me or do you think I am not doing appropriate work?" This is the first phase of performance measurement. There are three.

In my experience, there are stages people go through when first being introduced to performance metrics. They are usually similar to the following:

- **Stage 1** — "It's not me you need to measure, its him/her." (Points finger at another employee.) This is probably the denial stage.

- **Stage 2** — "Okay, I guess you (management) are not going to forget this new idea. (With sort of a disgusted tone.) What was it you wanted measured anyway?" This is the acknowledgment stage — it is real. "It's going to happen and there isn't much we (the employees) are going to do about it."
- **Stage 3** — "I have been measuring this and I found that there are some other diagnostic indicators that will help in the elimination of the root cause of the variation." This is the exciting stage — the stage of discovery.

Obviously, education plays a part in this transition, as does practice. Management must send a clear and consistent message that the measures are about process and not people. This is done through both actions and words. Management's signal cannot be *delegated* to the staff. The employees watch and know what management thinks is important! People always do what the norm of expectation is.

> Management's signal cannot be *delegated* to the staff. The employees watch and know what management thinks is important!

The most successful metric process start-ups involve education and communication. Normally, management starts the process, and once the metric performance percentages are confirmed from a standpoint of data and process ownership, the numbers are made visible. Some of the first education can come from simple communication sessions with the masses.

WEEKLY PERFORMANCE REVIEW

The weekly performance review was covered in detail in Chapter 14. Nonetheless, it needs to be reconfirmed that this is the king of metric management systems. No discussion around Class A metrics would exclude the mention of the weekly performance review. In high-performance organizations, the metrics continue to grow past the first-pass metric requirements for Class A. As diagnostic measures as well as additional barometric measures are added to the deck, they need to be part of the reporting process at the weekly process review.

POSTING THE METRICS

Once the Class A metrics are in place, performance numbers should be visible to the organization. Class A criteria specifically call for the Class A metrics to

Process	Process Owner	J	F	M	A	M	J	J	A	S	O	N	D
Business Plan													
Demand Plan													
Operations Plan													
Master Sched.													
Materials Plan													
Schedule Stability													
Inv. Accuracy													
BOM Accuracy													
Procurement													
Shop Floor Cont													
First Time Quality													
Customer Service													

Figure 16.4. Class A ERP Performance Board.

be posted on a visible board. Many organizations hang multiple copies in the factory and office areas. The boards should look like Figure 16.4. Note that safety is not posted on this performance board. Safety is normally posted separately and is not posted as a percentage.

Your board may look different if you add or alter metric labels. The important part is visibility. By making the metrics visible, awareness is increased. A good addition to the performance board hanging in the facility is a book of process metric descriptions. This is often just a multipage, stapled document hanging from the board on a string that allows people to look up the definitions of specific metrics they may have forgotten. The descriptions should include:

■ Process owner to contact with questions
■ The definition of the metric
■ Description of where the data come from
■ Copy of the calculation
■ Objective of the measurement

Metrics are the lifeblood of a high-performance organization. The best companies use them for competitive advantage. There are several things that are important for successful ERP implementation. Chapter 18 deals with the implementation methodology for efficient and successful Class A ERP achievement.

17

A WORD ABOUT
ERP SOFTWARE

Software is a major topic in most ERP discussions. Notice that we are well into this book on Class A ERP with almost no discussion concerning the software tools. APICS now actually has two definitions for ERP and includes the software as one version of what ERP is. Without software, ERP is impossible to accomplish in today's complex and competitive world. Software is like the tools in a carpenter's or cabinetmaker's bag. Elaborate cabinets could be made with no tools or with just a hammer and a chisel, but the time and efficiency of this process would make it undesirable and the skill-level requirements would be exponentially above the cabinetmaker who uses only modern tools. Instead, today's competitive carpenters and cabinetmakers use elaborate fixtures and powerful tools. Intricate designs can be copied and duplicated with great speed.

ERP business systems today are much like the power tools. These new systems can make schedule changes happen quickly and easily through the use of "drop and drag" capabilities. Almost as soon as the change is updated in the system, modern systems can quickly re-"net"/recalculate the changes to requirements and connect directly to supplier signals communicated through the Internet. Try that without software!

Having now underlined the importance of software, it is equally important to underline the need for good process. Without solid and robust process design, the tools do not work. Just as undisciplined and unknowledgeable people could not make cabinets with modern tools, neither can undisciplined and unknowledgeable people succeed at Class A ERP. This goes to the spirit of Class A ERP. Class A ERP was designed, developed, and evolved for exactly this reason —

good discipline makes ERP techniques and good supply chain management both possible and probable. As Tim Frank, CEO of Grafco PET Packaging, once said to me, "Why wouldn't everybody want to do this?" I agree with him. Why not?

Picking software is not so unlike buying a car or other important investment. Books have been written in the past on the basics of picking out a business system, but most were written in a time when business systems were still evolving quickly and there was little expertise in the marketplace. That is not necessarily true today. While systems are still evolving and always will, technology is more the driver of speed now than functionality. Most systems can do much more than is required or even desired by high-performance businesses. According to some estimates, in a well-managed and -executed software implementation, less than 30 percent of a system's full capabilities are regularly used. (This is a good thing, by the way!) Software companies have to design packages to work in many environments including poorly managed ones. I have been a speaker at ERP user-group meetings and have been taken back by some of the questions that are addressed to the system providers. Some of the popular ERP systems available today have over 5,000 system switches on the front end of the software that give users variation on how to use the system. It is doubtful that there are many people, if any, who really understand what all 5,000 switches do or how they affect other switch decisions.

I have had good luck with simple software. The simpler, the better, providing it has all the necessary functionality. I also do not endorse any specific software as I have seen all of the more popular tools used successfully and well — even the ones with the 5,000 switches! The message here is this: When it comes to buying ERP software, if you like the people you are dealing with, you know you can trust them, and there is a better-than-average chance they are going to be around in a few years to continue their support, the decision itself will probably be low risk.

> When it comes to buying ERP software, if you like the people you are dealing with, you know you can trust them, and there is a better-than-average chance they are going to be around in a few years to continue their support, the decision itself will probably be low risk.

The risks continue to be greatly minimized if the process disciplines and the skills and knowledge in your business are up to Class A ERP standards. By following the principles in this book and implementing the Class A ERP standards, software will be the tool or enabler to allow good process and results to flourish efficiently.

When I first started doing public speaking on the topic of Class A in the 1980s, I often used the "golf story" to make my point. This story has been around a long time in the Class A world and I am not sure who started it. It goes like this: If I went golfing with a professional golfer and we swapped clubs, who would probably have the best round? After all, I would have his very high-value set of signature clubs and he would have my garage-sale set. The answer is not rocket science. We all understand that the professional's disciplines and knowledge of the game would easily overshadow the fancy clubs used by someone who only plays a couple of times a year. Software is exactly the same.

When choosing software, look for the following list of criteria:

- Company reputation
- Software used in other similar businesses successfully
- Query ease and flexibility
- Service and support (locality and reputation)
- Price does matter (there are some packages out there that do not cost an arm and a leg that do quite well)
- Capabilities around a short list of specifics that seem important to the management team in your business

The obvious requirements probably do not have to be listed, but include:

- Good master scheduling capabilities
- Internet communications capabilities for supply chain management
- Metrics capabilities
- Easy-to-use query capabilities
- Warehouse location balance capability
- Easy links to spreadsheet uploads and downloads
- User-friendly menus
- Quick shortcuts to other screens, shared drives, and Internet that carry part numbers or data screen to screen

Most of the big names in software have appropriate functionality. That is not normally the issue. The hype around industry-specific software is not the same issue it was just a few years ago. More and more companies are realizing that their business is not as different as they thought it was.

If you take nothing from this chapter except one thing, it should be the understanding that all of the popular ERP systems work. Use the list to remind yourself to look carefully. There is some emotion in the purchase, just like there is in purchasing a car. Buy the system that you have the most support for internally, implement the Class A ERP disciplines, and you will be successful.

18

CLASS A ERP IMPLEMENTATION

Having read this book up to this point, you might think, "This all makes sense, but how do I get this started in my company?" The answer will be spelled out in this chapter. Following the implementation methodology will guarantee successful implementation. It has been proven many times before. The steps are universally appropriate, correct, and always effective. Few things in life are that sure.

STEPS FOR SUCCESSFUL CLASS A ERP IMPLEMENTATION

Education of All Parties

To get everyone on the same track right from the beginning, there needs to be shared goals, a common glossary of terms, and a shared vision. This is best done through education. Do not think if this as teaching "old dogs new tricks." Instead, think of this as reorganizing thoughts we have all had before into a central and shared focus. It is sort of like light shining through a magnifying glass. When the whole team is on the same path, change can happen quickly. This is the goal — making it happen as quickly as possible while keeping the spirit of the goal clearly in front of the team.

Top-management education usually takes about six to eight hours and is normally done in one setting. In this session, there will be a lot of emphasis on top management's role for successful implementation. This would include

Top-management education usually takes about six to eight hours and is normally done in one setting. In this session, there will be a lot of emphasis on top management's role for successful implementation.

the sales and operations planning (S&OP) process roles, management system roles, and "walking the talk." It often is done by outside experts, simply because top-management people do not always listen well to insider direction. Also, by getting this education done early, those in top management can easily position themselves as the champions of this process implementation, which is a powerful signal to send to the organization.

If the Class A champion has been chosen at this point, he or she should sit in on this top-management training. More education and training will follow, but the next step is to define the players and core team for the implementation. Once the players are chosen, the process can gain steam.

Education of the rest of the team comes in waves. The second educational focus will come with the Class A core team. The core team is made up of process owners from the major processes within each facility or plant. Each plant will have its own core team headed by a Class A champion for that plant. Education of these teams can be done in several ways. The most effective and probably easiest way for the company is to have this education delivered by professionals who can not only explain the basics, but can also give examples of their experiences with other companies. This can be especially effective in the early stages as converts and advocates of Class A ERP methodology are won over. Content of this core training should include detailed education on topics matching the processes disciplines in the Class A ERP business model (see Figure 18.1).

Initial education (usually two days in duration) covers the entire gamut of ERP, but follow-up education for the more detailed courses is best done in smaller increments and can take weeks to complete. APICS courses, outsiders, and purchased videos can all play a part in this important information transfer.

All people in the company will eventually be introduced to the concepts of Class A. This lowest-level education is delivered most effectively by the managers of those being trained. This approach puts special emphasis on the management team's understanding of the basics of Class A ERP. It also sends the strong message that this is important and not just another "float in the parade," one that "might miss me if I stay quiet and ignore it." Again, purchased video education can be very helpful in this regard if using managers for facilitators, although outside education can offer an especially effective alternative.

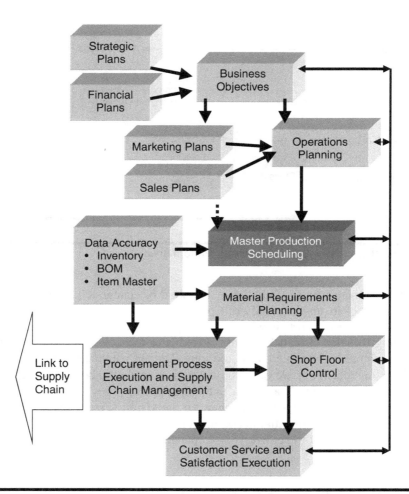

Figure 18.1.   ERP Business Model.

Establish Goals and Vision

The date for certification should be established during the top-management training session. In most businesses, the duration for achievement of certification is twelve months or less. It depends on the amount of change required and the internal appetite for it. This implementation can happen in a very short time. I have personally been involved in a couple that required only six months and many more that have required only nine months to achieve Class A sustainable

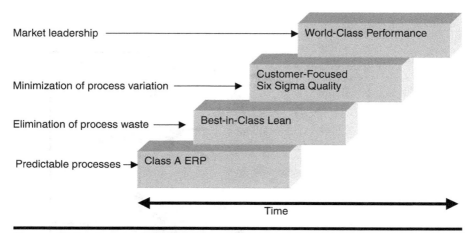

Figure 18.2. Journey to World-Class Performance.

ERP performance. In the "old days," this kind of change required months and months — not so, any more. My rule of thumb today is to plan for nine months. If the plan is too short, it will become apparent later. There is no harm in changing the plan if it makes sense. It is better to have a plan that is a little short in duration than one that is too long. Human nature leads to a tendency to delay real change if we have too much time to think about why we cannot do it or why we can always wait until tomorrow.

Use the journey steps to communicate the vision (Figure 18.2). The steps are not necessarily sequential in completion, but are sequential in start-up. This is often an easy way to get everyone to understand the pieces and how they fit together.

Organization Structure for Change

There are a few decisions that need to be made at the front end of a Class A ERP implementation. The short list is:

- **Name a company-wide Class A champion** — This person should be someone you can least afford to put in this job — yes, *least* afford. The job will include coaching, leading, and changing culture throughout the organization. This person must be able to communicate in the finance organization as well as within operations. The champion does not have to be an *expert* in everything, however, just a good communicator and leader. This is a desirable position for up-and-coming leaders. It is a fact

It is a fact that every successful Class A ERP champion I have worked with has impacted his or her career significantly in a positive way through this learning and experience.

that every successful Class A ERP champion I have worked with has impacted his or her career significantly in a positive way through this learning and experience. Typical Class A champions can come from operations, finance, or sales. Information technology lifers can be good choices, but some do not always adapt well depending on their real understanding of process as opposed to technology answers. Remember that the emphasis is on good process, not just good tools. Having said that, it needs to be stated that many information technology professionals have made great project managers. The point has to be made so as not to automatically pick the ERP software expert as the champion. Make the selection carefully.

■ **Officially name the members of the steering committee** — These are often the managers that report to the president or CEO. The steering committee will approve projects, review progress, make decisions when necessary, and generally give support for this major cultural shift.

■ **Determine the process owners for the S&OP** — This is not too difficult, but needs to be done formally. The operations plan is almost always owned by the VP of operations. The demand plan owner can be sales or marketing and is normally the VP of one or the other. Both sales and marketing need to be involved deeply.

■ **Develop a schedule for the S&OP meetings** — Out twelve months. Schedule this as "the first Monday of the month" or some similar time.

■ **Schedule steering committee meeting for the next nine months** — The reason for only nine months is that certification should happen within that period.

■ **Name the implementation team members** — These are the process owners at the plant level. If there is only one plant in your organization, this should be done up front. In a multiplant environment, it can wait at the plant level until after the communication has been done. People need to know what they are choosing process owners for.

■ **Finalize the kickoff date and how it will be communicated** — This is often done in a big way with much fanfare. It is an opportunity to educate the masses initially through communication. Using fanfare and top-management involvement will go a long way to convince employees that this one might be real. The kickoff needs to be closely followed with some evidence that the implementation has begun. This would normally include education, metrics, process owner assignment, etc.

Measurements and Data Collection

The measurements in Class A ERP are already determined. That eliminates a lot of wasted time debating (arguing) what they should be. Do not waste time getting them into place. Assign the process ownership and get the metrics in place — the quicker, the better. Top management must take the ball and visibly show ownership in the top-management metrics reviewed at the S&OP process meeting. A visible metrics board needs to be hung quickly in a conspicuous place. One plant used four huge white boards to show its metrics in oversized letters and numbers in the lunchroom. There was no missing it and the commitment to keeping it up was obvious. It showed the employee base that the company was serious.

Accountability and Management Systems

Every site should have a Class A champion responsible for the successful implementation and certification of Class A ERP. Normally, these Class A champions report on the dotted line to the overall Class A champion. It is important to pick the right leaders for this implementation. The typical characteristics for successful Class A implementation managers at both the overall and site level include: (1) deep product and process knowledge, (2) leadership characteristics, and (3) great respect within the organization.

Duties can vary, but in general the anticipated activities of Class A champions (Class A project managers) at the site level would be as follows:

1. **Facilitate education** — Class A brings with it a new level of discipline and accountability. Education will be delivered in many areas. Some will be done by staff, but local education will also be administered by process owners and supervisors.
2. **Develop project plan to cover all of the business model processes and required audit criteria for successful Class A certification** — The project manager will be responsible for managing the project plan for successful implementation of Class A.
3. **Oversee process ownership assignments** — Each of the metrics will have associated process ownership. This ownership normally will not reside with the project manager, but it is the responsibility of the Class A project manager to oversee the assignment and follow up as necessary for Class A levels of performance.
4. **Oversee implementation of metrics** — Design, data sources, and integrity of the metrics will be the responsibility of the project manager. Process owners will execute the metrics once determined and approved.

5. **Oversee population of performance metric visibility tools** — There will be visible tools available to publish performance results from the facilities. The Class A project manager will see that these tools are populated.

6. **Chair project review sessions on Class A implementation within the facility** — The overall Class A project champion is the project manager and owner of the Class A ERP implementation and certification. It should be thought of as a condition of employment that the facility and the company be certified as Class A ERP compliant.

7. **Aid in the development of reporting formats (including S&OP)** — The process of linking processes includes a lot of communication and schedule knowledge and knowledge transfer. The Class A ERP champion is the cog in the gear set for establishing the formats and practices for process linkage. Either do it or make sure it gets done.

8. **Become an ERP advocate within the business** — One of the biggest jobs of the Class A champion is to sell ideas. Class A is a logical process, but human nature will work against most change. The champion must be unwavering in his or her dedication to the Class A ERP cause as the process starts to get traction in the company. This would be evidenced by attending department meetings and speaking on the topic of Class A ERP, promoting Class A at company events, writing articles for the company newsletter, creating videos marketing the process concepts, and generally promoting Class A ERP.

> One of the biggest jobs of the Class A champion is to sell ideas.

9. **Reports progress to the Class A ERP steering committee and company Class A process management** — This reporting generally happens at least once a month and more frequently in the first few weeks. This reporting should be formal to ensure that homework is prepared for this monthly meeting. Suggest that the champion prepare a slide show each month with metrics, Gantt chart progress, and bullet statements of progress.

10. **Organize accountability reporting infrastructure elements (weekly performance review, project review, daily plan adherence review as required)** — The master scheduler is a main coordinator in much of this accountability, but the champion is responsible for making sure it is happening in a timely manner, especially in the beginning of the implementation. Once the pump is primed, the champion has a diminishing role in the management system. The actual management systems

that must be implemented for proper ERP success are S&OP, weekly clear-to-build, weekly performance review meeting, project review, and daily walk-around. Proper accountability is the key to successful management systems. Not to be confused with bloodletting, proper accountability means people are expected to show up with their background work completed and their facts gathered. There is no need for yelling and screaming, but there is a need for a choice — compliance or cost.

> There is no need for yelling and screaming, but there is a need for a choice — compliance or cost.

11. **Determine/audit internal data sources for required reporting** — Some people will try to reinvent their metrics as they are sure that their way is the better way. If this is not well communicated and agreed to by the steering committee, it can go unnoticed and ultimately deceives management into thinking something is different than it really is. It is important to know what the metrics are measuring and how the metric works. Documentation can help this, especially if it is audited periodically. All changes to metrics must be approved.

12. **Coach the facility to successful certification by goal date** — The bottom line is that the champion is accountable for Class A certification by the goal date.

13. **Celebrate and publish successes** — Celebrations are left out of the plans all too often. The problem is that the more aggressively companies go at process improvement, the higher the expectations become. This often makes goal attainment anticlimactic. Be aware of that issue and avoid it by scheduling and executing celebrations at appropriate times. These do not have to be elaborate. A cake and a two-minute speech is sometimes all that is necessary to get the word out. Celebrations are a form of education. They help teach the organization what good behavior is.

Documentation

Once processes are determined and discipline requirements are designed, documentation must be completed. This is the final chapter in the implementation. Doing documentation gives the organization a chance for sustainability. Once the documentation is completed for the first set of processes, auditing should begin. This is normally done by the comptroller's office, just as inventory accuracy would be audited for the asset register. The International Organization for Standardization (ISO) has a good expectation for documentation auditing, and Class A ERP process documentation should go into the ISO-controlled documentation as soon as it is written and approved.

Training for Sustainability

The documentation sets the stage for training materials required for ongoing, sustained process predictability and performance. Training responsibilities need to be part of process ownership, as does backup planning for succession and absenteeism needs.

Ratcheting of the Goals When Appropriate

As the metric performance is met, the expectations need to be ratcheted up. It makes no sense to have process owners without goals for improvement. Keep the objectives realistic but continually improving in final goals and expectations. For the sake of celebration and achievement, goals generally should not be ratcheted until the existing objectives are met and are proven to be sustainable for a few weeks.

SUMMARY

Class A ERP is the first step in the journey to world-class excellence. Much can be gained in competitive advantage by focusing on the goal of process predictability, schedule adherence, and data accuracy. It is not the end-all or even close to it. It is just the first step. The objective of this book was to present a lot of evidence to show that integration of ERP with lean and Six Sigma not only makes sense, but is important to get the most from your continuous improvement efforts. There have been several references throughout the book to both lean focus and Six Sigma methodology. To further make the point, the next chapter will be dedicated to this integration, just in case the message went unappreciated in preceding chapters.

19

CLASS A ERP INTEGRATION WITH LEAN AND SIX SIGMA

Lean and Six Sigma have been mentioned several times throughout this book, but there is still a lot to cover for the few people who are still convinced that there is an integration issue among the three elements of Class A ERP, best-in-class lean, and customer-focused Six Sigma quality (see Figure 19.1).

BEST-IN-CLASS LEAN

The best place to start is with an introduction to lean and the methodology behind the label. Lean strategy is focused on the elimination of waste. Some will argue that process design must link to customer needs and that elimination of waste is only defined after the customer requirements are well understood. At this point, lean experts will immediately argue that lean thinking is all about customer requirements. This is not something worth arguing about because at the end of the day, there are more similarities than differences between Class A ERP, best-in-class lean, and customer-focused Six Sigma quality.

> This is not something worth arguing about because at the end of the day, there are more similarities than differences between Class A ERP, best-in-class lean, and customer-focused Six Sigma quality.

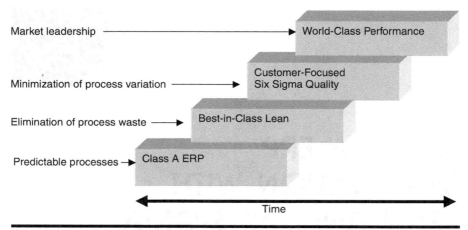

Figure 19.1.   Journey to World-Class Performance.

The whole idea behind separating the approaches is simply to define the journey into bite-size pieces and celebration points.

1. **Becoming predictable** — Making promises. When meeting market requirements, customer service promises can make a big difference in competitive advantage, but they are not the whole picture. By focusing on process capability first, processes are reinvented, redesigned, or made repeatable through the use of metrics, management systems, and root cause analysis and barrier elimination. The deliverables from this element of process improvement are the management system and the cultural shift to measurements and analysis. With the Class A ERP emphasis on sales and operations planning (S&OP) and supply chain management, there is little need to do prioritization of projects. Because the methodology is well defined, the process becomes a very efficient cultural-shifting strategy. It is not the whole answer, however.

2. **Improving the processes by eliminating waste** — Lean focus works well for the second step in the journey to excellence as the emphasis naturally shifts from process design and capability to process improvement. Admittedly, there are many examples where the processes get redesigned a second time in this space. To think that there is ever an end to process improvement is ludicrous anyway. This may not be the last time a single process gets completely redesigned. The specific lean focus is different, however. The emphasis in this space shifts from process capability only to process flexibility, speed, and responsiveness to customer need, a step above the Class A ERP emphasis. The ERP prereq-

uisite deliverables of the daily/weekly/monthly management system and habits of measurement also make lean very efficient as the emphasis moves to speed. At the celebration point of best-in-class lean, companies can expect fast and responsive processes with measurements and management systems that will continue the improvement into the Six Sigma space.

3. **Lowering costs further by minimizing process variation** — All the steps to world-class excellence share a common thread: continuous improvement driven from understanding and eliminating root causes of barriers to achieving ever-increasing high performance. The third step in some businesses is the step that never ends, Six Sigma. Process variation in all organizations is unavoidable, but in high-performance organizations, *decreasing* process variation is also unavoidable. In statistical terms, one sigma is equal to one standard deviation. Recall from college statistics that as process variation is decreased, so is the standard deviation. Customer requirements are the limits of acceptability in any process, and as the standard deviation is decreased, the likelihood of missing customer requirements is also decreased as long as the average of the process is located within the upper and lower specification ranges. This is the theory behind the Six Sigma capability measurement. If the process allows six standard deviations between the mean and the customer limits, the likelihood of a defect is almost unheard of — 3.4 defects within 1,000,000 opportunities (Figure 19.2).

Best-in-class lean, like Class A ERP and Six Sigma, is both a methodology and a goal of performance. With the focus almost entirely on elimination of waste and increasing flexibility and speed, lean thinking starts from the premise that there is waste all around us. Waste is generally found in the eight different

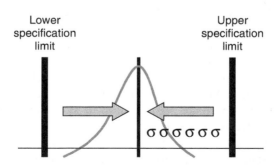

Figure 19.2.   Visible Description of Six Sigma.

areas that follow. Look at these one at a time and apply the logic of lean integration with Class A ERP; it really is a very logical fit.

1. **Unnecessary material movement** — The first topic of waste is found in every corner of almost every business. If we start with the premise that *any* material movement is an opportunity for elimination of movement, a lot of opportunity arises. In one business I worked with, there was an assembly line like so many others, but in this particular environment, prior to Class A influences, schedule adherence was measured by material movement *into* the line rather than counting completions for the metric. Since this was not Class A compliant, the measurement was changed to units finished. As they worked to get the order daily completion/schedule adherence metric to standards, what they found out was that the line was about sixteen feet longer than it needed to be for optimum efficiency and schedule adherence. It was the Class A measurement that led to this finding — true integration between Class A ERP and lean. If capacity is at a premium and Class A metrics are in place, material movement opportunities will be flushed out every time as the result of metrics driving root cause analysis and resulting actions.

2. **Unnecessary worker movement** — One time, a plant manager wanted me to see the improvements they had made to their material flow since my last visit. When we got up to the second floor where the assembly operation was executed, I could see no difference from the last visit. He then proceeded to show me how they had moved a machine approximately six inches! It was not even noticeable to the outside observer, but they had realized that this movement had affected their ability to meet their tight Class A ERP schedule requirements. Along with safety requirements, that driver led to the machine shift, which eliminated a step that the operator had to take between operations in the work cell.

3. **Wasted space** — At The Raymond Corporation, growth was a constant. Eventually, we subcontracted operations to gain capacity, but in the early years of my career, we tended to think we had to do everything ourselves, and with every growth in sales and new products came the shifting of machinery and work cells to gain capacity. Space was at a premium all the time. Also in most businesses today, inventory turns are a critical measurement at the monthly S&OP review. Investments in manufacturing space are not best utilized by piling inventory in it. It is not very difficult to see the relationship between the lean concept of eliminating wasted space and Class A ERP performance requirements.

4. **Wasted time** — A major focus of lean, wasted time is one of the first elements of waste identified in Class A implementation. Rate or uptime

is a normal Class A ERP metric and drives root cause analysis toward this area of waste.

5. **Wasted material** — Scrap is a frequent issue when driving inventory balance accuracy. It is often not accounted for correctly and results in inflated balances and inventory shrink. Class A data accuracy drives this to the surface very early in the process. Rework is another similar situation resulting in material being wasted along with labor. Additionally, lost or damaged material does not go unnoticed when Class A ERP measurements are focused on high levels of data integrity.

6. **Wasted effort** — Wasted effort (such as in poor quality) is quickly identified through the first-time quality Class A requirement. Process design quality is one area that gets special attention through the bill of material and standards review requirements in Class A ERP.

7. **Wasted knowledge or information** — Using information or lessons learned from others is an especially efficient use of resources. Team members should read and be active in professional societies. Networking within these organizations is very healthy. Do not ignore or waste information in the real world.

8. **Wasted creativity** — Wasted creativity has always been one of my favorite waste opportunities. When we engage the entire workforce, it becomes enjoyable to see ideas generated. In one bakery I worked with several years ago, a third-shift oven operator came up with some new ideas (including pumpkin-flavored cake cones) that had not been thought of by management. Although the idea did not end up in stocked product, the idea did generate other avenues actually pursued by the company. Creativity comes in many forms. Some of the best ideas I have seen implemented were not thought of by management or engineering, but by people from the office or factory floor. Engaging the entire workforce (from the head down — not just their backs) is a big step forward in asset utilization and another clear integration with Class A ERP and lean.

LEAN TOOLS

It is not the intention of this book to be a guide to lean. There are other books that do that. It is the intention of this book to show how lean easily and logically integrates with Class A ERP. Class A ERP is all about the introduction of metrics as a main driver of action and analysis in the business. Without good problem-solving tools, the actions would not be a result from the metrics. It is the analysis that leads the way. Lean tools are at the center of this analysis. As

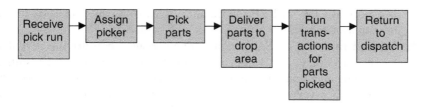

Figure 19.3. Process Map Example.

explained in Chapter 13, process mapping are an unavoidable tool in a high-performance organization. These various mapping tools include several important methods to identify waste in a process.

In a Class A process, after analysis and improvement are at the 95 percent capability level, the expectation is that this process (95 percent of the time) will result in acceptable performance. For example, in inventory accuracy, the Class A process will establish that 95 percent of the inventory location balances will be 100 percent accurate at any one time. This does not necessarily mean, however, that the process is without waste. The second-level look at this process may find that there is an opportunity to eliminate moves or transactions or even inventory. This second-level look is often the look through a lean lens.

Process mapping allows this process to be viewed for its elements (see Figure 19.3). During the inventory accuracy investigation and improvement activity, it might be determined that there was some discipline missing in the pick operation. Figure 19.3 might be the result of the documented map.

After the process is repeatable to a Class A performance level, the next level of improvement must take place. That level would most likely start with a lean view of the process. The process map is the perfect place to start. In this example, after scrutiny and asking the question, "Are each of these steps value-add?" the resulting process could be much different (see Figure 19.4).

Without knowing the specifics, this could be the right and appropriate next step in the improvement process of eliminating waste. Clearly, lean is playing an important role in the continuous improvement process for this plant. The job

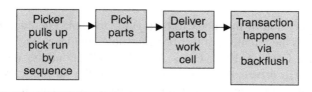

Figure 19.4. Leaned Process.

would not be done; as in this example, there are still opportunities for waste elimination as parts are still being stored in a pick location and not right on the line. The job of continuous improvement is never done — never!

CUSTOMER-FOCUSED QUALITY (SIX SIGMA)

After the lean celebration, the next performance level goal that is extremely high and one that is seldom, if ever, totally reached is the Six Sigma objective and methodology. With only 3.4 defects allowed in a million opportunities, the standards are higher than most objectives in the past and certainly higher than in Class A or best-in-class lean. Class A ERP is the right foundation for this third level of performance criteria and tool set. Gardiner Arthur, an NCR master black belt and friend from Scotland, has always argued with me that Class A ERP should use a more Six Sigma–related approach to metrics by using variation-based metrics as opposed to strictly percentage performance metrics. Gardiner makes a good point. Variation-based metrics are superior to percentage performance metrics as they define the swings in variation — how far out of spec the percentage of misses go. When just using percentage measurements, one can only see that some of the process results are out of spec, not how much these results are off. This is part of the concept behind the step approach. Class A has many focus areas. To help keep the emphasis where it needs to be, Class A says let's get the process to a level where at least 95 percent of the time we are doing what we plan. This is a major accomplishment in most businesses I visit. When this is accomplished, it is time to tweak these processes to even better levels of predictability. Class A has a 95 percent threshold; Six Sigma has a 99.99966 percent threshold. In my experience, there is value in taking this ratcheting in steps. As one understands both methodologies, it is also apparent that there are no conflicts between the two approaches. The integration is a natural one and is logical.

Six Sigma is actually a definition of the levels of defects found in a process. As stated earlier, Six Sigma means there are predictably 3.4 defects for every million opportunities in a given process. In reality, Six Sigma has also become a label for a specific methodology of problem solving and project management. According to many Six Sigma experts, the methodology label was started by Motorola and made famous by GE. Jack Welch, former CEO of GE, was one of the biggest advocates for this methodology and helped to make it popular at many companies. Most large companies have some form of Six Sigma process improvement methodology in process. Some of the other companies that utilize Six Sigma somewhere in their strategy are Dell, Lockheed Martin, Honeywell (and former AlliedSignal), NCR Corporation, and hundreds of others.

All processes have variation, so a methodology to minimize it would make a lot of sense to most intelligent people. Few processes are better because of the variation; 3.4 allowable defects in a million opportunities, to some, might seem like a high bar and it is. That is why I became an advocate of the stepped performance improvement approach (Class A ERP, lean, Six Sigma). Most find that Class A is not a cakewalk and the performance criteria allow 5 percent noncompliance. In Six Sigma terms, that is 50,000 allowable defects in a million opportunities or approximately 3.2 sigma. Now, all of a sudden, Class A ERP sounds like it is pretty low quality. Remember that Class A is a celebration point on the way to world-class performance. You will find it is a valid one as well. Ninety-five percent or 3.2 sigma is more difficult than one might think. Nonetheless, I hope that the case for integration is an easy one to digest. As stated, all processes have variation. That is as true in the office as in the factory. Some of the best opportunities are in areas like payables or strategic planning. Six Sigma has a wide berth of topics; it really has no boundaries.

I do work for the State University of New York's Empire State College in my spare time. My role there is to verify experiential learning for college credit. I recently interviewed a student with several years at a reputable defense contractor in the United States. She was from finance and was working on her black belt certification. It is always good to hear that finance people are getting involved in process improvement and change. I do not want to pigeonhole anybody, but the accounting profession has been locked into a specific approach for many years — accounting standards have created a culture in some organizations that makes change difficult. It was refreshing to listen to this finance professional talk about the need for change. I could not agree more.

In most companies, Six Sigma is built around the DMAIC (Define, Measure, Analyze, Improve, Control) methodology (Figure 19.5). Those who are Deming fans will see the similarities between the DMAIC process and Deming's Can-Do-Check-Act wheel. The approaches are based in the same learning.

DMAIC was described in detail in Chapter 13. In this chapter, we will take the DMAIC process to another level of detail and describe what each step might look like and what tools would be properly utilized.

D — Defining the Problem and the Tools to Use

One of my favorite tools to use in defining a problem or opportunity is called a SIPOC. SIPOC is an acronym that stands for suppliers, inputs, process, outputs, and customers. The SIPOC is a great step to better view a process for the potential influence on improvements to the process (see Figure 19.6).

It is important to know the details behind a process as the project is launched. By asking about suppliers and sources of input to the process, opportunities are

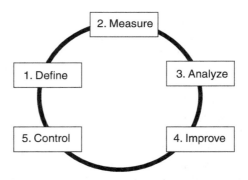

Figure 19.5. Six Sigma DMAIC Approach.

This is the spirit of Six Sigma — opening up all the possibilities to make sure the best solutions are found.

opened up for solution sources that might otherwise be ignored. This is the spirit of Six Sigma — opening up all the possibilities to make sure the best solutions are found.

Outputs are just as important. When mapping the process, it is often discovered that data from the process are used by people unknown to the process owners. The reverse is also sometimes true: people supply data or outputs only to find that they are not used or needed any more. Asking the questions posed by the SIPOC is a good exercise. Other tools used during the Defining stage include the project charter, brainstorming, process mapping, and stakeholder analysis.

Stakeholder analysis can be fun for most teams. In this exercise, team members are asked to define the people who will have the most influence on this process as change is introduced. This can mean, for example, that if one member of the leadership team is very interested in this particular process and will have specific inputs and expectations, this person will be defined as a major

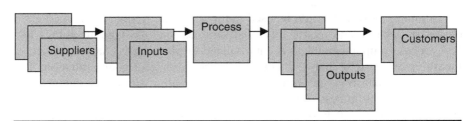

Figure 19.6. SIPOC.

stakeholder, even if not officially assigned to the team. Stakeholder analysis includes defining not only level of interest, but also the influence factor, both negative and positive. A client recently had to replace the VP of operations because this person had become so set in his ways that change, as determined by a continuous improvement process within the operations group, was being dragged down by his lack of engagement and enthusiasm. He was not necessarily opposing the implementation of Class A ERP and lean, but just was not promoting it, which, in itself, sent a powerful message. This type of influence is good to get out on the table in the beginning of the project. Define who the team needs from a support standpoint and who the team has. It is not always the same list of people. Ranking influence can be helpful in determining actions. Sometimes education and marketing of ideas are part of the solution.

The deliverable from this stage is an approved project charter or scope document that includes things like what the objective is, how long the team has to finish the project, who the team members are, and even what is *not* included in or is out of scope for the project.

M — Measure and Collect Data

This stage is easily understood, but not always done with a thorough eye. There are tools that can make this stage easier and more effective. Probably one of the most widely used is process mapping, the documentation of a process by visibly depicting the components or actions of the process. Again, this can be very helpful in determining the opportunities and possible solutions. Process mapping was covered in detail in Chapter 13.

Another valuable tool in the "measurement and collection of data" stage is the fishbone or cause and effect diagram. This is a tool that reminds the user to look in all the corners for opportunities. The reminder is normally the "6-M" memory jogger: manpower (people involved), materials (both materials used in the process and chemicals used in the periphery), methods (or process approach), machinery (tools or equipment used in the process), measurement (how the data are displayed or categorized), and Mother Nature (things like barometric pressure and altitude above sea level, humidity, temperature, etc.).

Common tools that would be used in this stage are frequency charts, Pareto charts, run charts, and metrics. The deliverables for this stage are things like key process output variables, a data collection plan, actual data collected, a measurement system analysis, and many times, a project goal validation. The measurement phase is the point at which the baseline sigma is calculated for the improvement process. This stage does not necessarily end as the next stage starts. Collection of data often goes on well into and even after the project is completed.

A — Analyze the Data for Opportunities and Possible Solutions

Analysis is a key component of Six Sigma. I often see teams jump to solutions too quickly. They are in the "let's try this" mode. Six Sigma requires deeper thinking and understanding in terms of the causes and effects within the process in question. Some of the tools normally used in this stage are similar to those used in the "M" (measure) stage including fishbone analysis, process mapping, and CEDAC (cause and effect diagram with the addition of cards). Another important point in this stage is to have the finance people involved in the analysis. It is especially important for the Six Sigma process to have the savings or impact verified by the financial experts in the business. The finance people should be involved in each project to a level to understand the potential financial impact of a project. Financial management involvement is not limited to this Analyze phase. Finance should be involved any time data indicate reasonable assumptions of payback from the project.

There are different types of financial impact. The most beneficial are obviously projects that result in direct benefits that decrease existing costs and improve cash flow. There are also benefits that would be categorized as cost avoidance. These savings do not reduce budgets or improve cash flow, but can be an improvement to planned cash flow. The Analyze stage deliverables include process maps, key process input variables, proposed solutions, and financial analysis reports.

I — Improve the Process

The fourth stage, Improve, is a focus on implementing the solutions determined in the Analyze stage. This does not mean that the ideas will always work. A testing process that determines the best solutions is appropriate and should be documented for future reference. Sometimes it makes sense to pilot solutions. Other times, confidence dictates full solution implementation. The deliverables from this stage are the proven solutions and verified financial results. Measurements need to be continued well into the "I" stage to ensure that improvements have had the process impact planned. At this stage, the solutions normally have high confidence levels and are becoming the standard process, replacing prior versions. After implementation of the new process improvements, a new sigma capability should be taken to validate the improvements.

Probably the most common tool in the Improve stage is the Gantt chart. The Gantt chart shows the actions of each expectation and the times for starting and completing various actions. In most cases, there are critical path elements that include prerequisites to some important steps. Project management software can be helpful in tracking these steps.

C — Control

The last step in the DMAIC process is the step to ensure sustainability. This step is normally focused on fool proofing, mistake proofing, and documentation. Documentation is especially important. Many people are proud about not wanting to take time to document and will tell others that paperwork annoys them. That is a popular view. We all hate paperwork, but I have found through experience that if a process is to be enforced, the rules must be documented. It is impossible to enforce disciplines that do not exist except within people's minds. Deliverables in this space include the control plan, policies, standard operating procedures, and work instructions. It is powerful to have strong documentation expectations and audits to make sure the documented processes are followed, providing the documentation is in sync with the performance expectations of the business. Deliverables should also include celebrations, documented savings and controlled process improvements, and final project closure.

OTHER SIX SIGMA TOOLS

A discussion on additional Six Sigma tools probably would start with variation-based metrics. Nothing is more important in any process improvement methodology than measurements, and within Six Sigma that is no exception. Variation-based metrics are not written as percentages. Some Six Sigma training will actually make a case against using percentages in any application. Variation-based measurements are always visibly shown with calculated limits (see Figure 19.7).

Belt Recognition in Six Sigma

Many readers have heard or read about green belts and black belts earned through Six Sigma efforts. Although there is variation from one implementation to another, generally the GE standard is the one most often referenced. The following examples are not meant to document any specific company's policies, but instead are a general guideline. Most green belt recognition standards include:

- One to two weeks (forty to eighty hours) of training and education on Six Sigma methodology and tools
- One project as a team member with a minimum savings threshold
- One project as the project leader using proper Six Sigma tools
- Present the project results to a Six Sigma council for certification

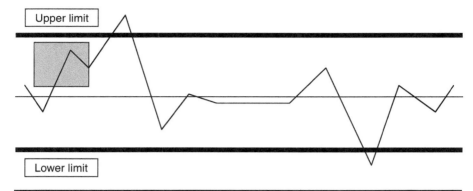

Figure 19.7.   Variation-Based Metrics.

Black belt minimums often call for additional requirements to be met:

- An additional week (forty hours) of education and training
- Facilitate a minimum of six successful projects
- Present the project in front of the Six Sigma council for certification
- Pass a black belt certification test

Six Sigma "belt" certification is a clever way to promote the culture shift and get people on board for the new standards of process improvement. People are motivated to get the recognition. In most successful organizations using Six Sigma, top management follows the pattern by becoming certified. It clearly sends the right signal to the organization.

Six Sigma is a logical and rigorous methodology to help an organization reach the highest levels of process proficiency. There is no conflict with Class A ERP or lean; in fact, the processes complement each other. Six Sigma is a high standard and one that can become a great motivational process. The celebration points in Six Sigma are less defined than in Class A, allowing an unlimited focus on goals. Defects are defined by the customer specification, and Six Sigma program performance is determined by the Six Sigma council, normally the top-management staff. Integration with Class A and lean is accomplished simply through a sequential evolution of standards and focus. Six Sigma is a welcomed performance level in the journey to excellence. It is so high a level that the next performance plateau is almost hard to imagine.

20

THE ROLE OF CONSULTANTS IN CLASS A ERP

The last topic to be covered in this book has to do with obtaining outside help. The world of consulting is made up of all types and approaches. The approaches vary from "do it for you" to "don't really do anything." The best results normally come from something in between. When outside help is hired to design and implement solutions, the result is often a lack of inside ownership of the solutions. Lack of inside ownership is a recipe for failure. There are some parts of Class A that best utilize outside experts. These areas include:

- **Education of top management** — Outside experts, people who have been through the ERP implementation before, normally have battle scars that increase wisdom on the topic. The reason many ERP implementations fail is almost always due to top-management issues. Outside experts who have the respect of top management can point out the required role of management and even the shortcomings of the existing process without the same stigma that an internal resource might fight. It is also extremely important to have the top-management team and the workforce on the same page.
- **Education of the core team** — The core team ultimately has the accountability for the success of the ERP implementation. Knowing about pitfalls before they happen is an asset, and education from outside experts with experience gained from several implementations can be a

big help. These resources can provide insight into specific activities that lead to quicker results. This is a good use of outside help.

- **Providing materials for critical mass education** — Materials can come from many sources: outside experts, the American Production and Inventory Control Society (APICS), local user groups, colleges, etc. I have had good luck with digital video materials for the detailed self-facilitated ERP and supply chain materials. Developing these materials internally would take months and many resources, and then would probably not be up to the quality of materials available on the market. Digital versions have made these products more affordable, especially when shared via a company server. Agreements can be purchased for this distribution from some suppliers.

- **Defining the Class A model** — There are a lot of models out there, and most are different versions of the same process. Inventing one would be risky. It is better to start out right than to have to backtrack. The entire team needs to center on the same glossary of terms. The choice is to debate it or choose a supplier that prescribes a specific version and go with it.

- **Coaching the process going forward** — Outside experts can help coach both up and down. My first consulting experience in the ERP space (called MRP II in those days, 1987) was as a practitioner hiring a consultant. As the arrangement developed, I would occasionally suggest topics to comment on in the consulting report. It is often difficult to point out opportunities for top management to play a different role in the implementation (or sometimes even "get their act together!"). This is a natural role for outside experts. Top management will listen to them even when the inside information tells the same story. This is a sad, but true, reality and one we are all aware of.

- **Providing examples of project plans** — Outside experts have examples of previous implementations and can help with networking as well. This can save time and effort.

- **Providing network opportunities with other companies in the same situation** — Being able to contact other companies that have been through this before can be helpful. This can also happen with contacts through APICS networking.

PAYBACK

The cost of consultants in the ERP field, like other fields, can be pricey. Many of the better ones can return the investment in implementation efficiencies and quicker returns in short periods. One quick return in many businesses is the simple act of eliminating the physical inventory. In large businesses, this can cost hundreds of thousands of dollars annually and is taken for granted. Class A requirements of data accuracy always result in the elimination of the physical inventory — always. The only exception is if you are in a country that requires an annual physical inventory to be taken by law. Some companies have already done the work to eliminate their annual counts, however.

The payback comes from successful implementation. Using experts simply helps the process happen faster. Areas of payback expected from a Class A ERP implementation include improvements in schedule stability leading to less expediting support required, less priority freight, fewer missed shipments to customers, and less inventory required as plans become more robust. The payback is linked directly to the top-management dedication to the culture shift and can be enormous depending on the starting point. It is not unusual to see returns of 200 to 300 percent in the first year, making the process easily self-funding.

Whether you use outside experts or not, the keeper of the body of knowledge is the APICS organization, and it is recommended that APICS be part of the strategy. This organization can provide good materials and often can inspire behavior changes as key employees attend conferences. I am especially supportive of the APICS International Convention held annually in North America. A couple of years ago, I attended an APICS seminar in Nashville with a friend. After listening to the featured speakers, my friend went back to the top management at his firm and convinced them to initiate an improvement process that included Class A ERP as the starting point. Today that effort is well into the lean phases and has seen many benefits from the process.

Without outside exposure, breathing your own air for years makes the business environment a bit incestuous. Not to offend anyone with that statement, but it is true. It is simply not healthy for an organization to exist without benchmarking with other markets or seeing how others are experimenting with process improvement. It is okay to borrow good ideas; as a manager in business, it is your duty to do so. I am not advocating breaking any patent or copyright laws, but I am saying that managers need to read, watch, network, attend classes, and generally seek out better methods both from self-experimentation and through education and training. It just makes good sense.

Class A ERP is the foundation of good performance in both distribution and manufacturing. It is logical, proven, and, when done correctly, easily accomplished. Hanging a Class A plaque as official certification does not gain market share in itself, but it does create a shared goal that the business can focus on, improve with, and never be wrong in doing so. Whatever you decide to call your improvement process, best wishes on your journey. E-mail me at info@sheldoninc.com and tell me about your wins. God bless!

APPENDIX A: SUMMARY OF CLASS A ERP CERTIFICATION CRITERIA

The following is a summary of criteria for the Class A certification process. It is not intended to document the full certification criteria in detail.

PRIORITIZATION AND MANAGEMENT OF BUSINESS OBJECTIVES

Strategic planning process	An annual strategic planning process exists and is managed by the CEO.
Business imperatives	A short list of twelve-month measurable objectives is in place.
Project funnel	A repeatable process for developing and collecting project ideas is established.
Prioritization of projects	A repeatable process for prioritizing projects exists and is linked to the business planning process.

Resources and skills required	A talent review process exists.
Review process	Management predictably reviews projects and is involved in the empowerment of new resource assignments.

SALES AND OPERATIONS PLANNING (S&OP) PROCESS

Process design includes demand and production plans reviewed monthly by top management	Demand forecasts and production plans as well as inventory and financial plans are distributed to management for review in a twelve-month rolling format.
Meetings	The S&OP meetings are held every month at the same time and are scheduled twelve months in advance. They are attended by top management.
Performance	Demand accuracy is 90+ percent accurate. Financial and production plans are 95+ percent accurate. All top-management plans are measured in consistent product families.

HUMAN CAPITAL MANAGEMENT AND INVESTMENT

Professional society affiliations	Organizations such as the American Society for Quality Control (ASQ) and American Production and Inventory Control Society (APICS) are supported.
In-house education	Education is provided for existing and new employees.
Tuition aid programs and guidelines	Outside education is supported.
Communication meetings	Communication meetings are held regularly with departments, divisions, plants, and the company.
New employees training	New employees are given a full introduction to improvement processes and work processes.

Existing workforce training	All employees are regularly updated with appropriate training for improvement and problem solving.
Skills assessment	A regular skills assessment is done to understand skills needs and plan for filling gaps.
Class A training	Class A training is provided for all employees.

NEW PRODUCT INTRODUCTION

A gated process exists for introducing new products	There is a document describing the process. Deliverables and process ownership are defined.
Discipline is obvious, as are measurements and follow-up	The gate meetings are held with top-management participation.
Ownership in new product introduction is obvious	Accountability is obvious through schedule measurements and process ownership, including demand, supply, and engineering functions.

MANAGEMENT SYSTEMS

Yearly	Strategic planning process, talent management, and succession planning all exist and are predictable processes.
Monthly	The following management systems exist: S&OP, project management.
Weekly	Weekly performance review, clear-to-build.
Daily	Daily walk-around and visible factory schedule boards exist.

SCHEDULING DISCIPLINES AND PRODUCTION PLANNING

| Master production schedule (MPS) | The MPS is aligned with reality at least once a week. |
| Rules of engagement | Rules are defined and followed concerning lead times, available to promise, transit times, etc. to customers. |

ABC stratification

Inventory is managed through stratification of criticality.

Inventory management

Excessive inventory is managed regularly, obsolete inventory is dealt with, and turns are measured, reviewed, and appropriate for the market.

Performance

MPS accuracy is 95 percent.

DATA INTEGRITY

Inventory

Inventory cycle counts are done daily.

Location as well as process are separated

Raw, work in process, and finished goods are measured separately. Each has location balances within the respective area.

Bill of material (BOM)

BOM audits are done regularly. New product introduction audits include postproduction BOM accuracy audits.

Item master

Item master fields such as lead time and cost standards are audited for accuracy.

Engineering change

BOM revisions are tracked and documented. Special documented authorization is required to deviate from the BOM.

Performance

Inventory location balance accuracy is 95+ percent accurate. BOM accuracy is 98+ percent accurate.

Security

The business system has appropriate system security and is reviewed and updated regularly.

EXECUTION OF PLANS

Supply chain management includes elements of supply chain partnership

Information sharing is extensive and seamless with critical suppliers.

Shop floor control is linked directly to the MPS

Daily reviews are done for performance to the MPS.

Performance

Both procurement and shop floor control are at 95+ percent schedule adherence.

MEASUREMENTS

All Class A measures are in place

Performance is 95 percent.

Measures have clear account-ability

Each process has clear process ownership by one individual.

Visibility of performance

Performance is visible in the form of obvious postings in the facility.

APPENDIX B: STEPS FOR SUCCESSFUL CLASS A ERP IMPLEMENTATION

The following list is a sequence of events for the successful implementation of Class A ERP performance.

Education	■ Top-management education on Class A ERP, sales and operations planning (S&OP), measurements, and management systems
	■ Core team and middle management education on all aspects of Class A ERP and lean
	■ Critical mass of both office and plant personnel education and training
Vision for continuous improvement	■ Class A date chosen and communicated to the masses
	■ Journey to world-class performance defined and communicated, including lean and Six Sigma focus
Organize for improvement	■ Steering committee defined and established
	■ Class A champion designated
	■ Core team empowered

	■ Process owners defined
	■ Outside resource secured
Measurements	■ Class A and lean metrics defined and documented
	■ Accountability assigned
	■ Measurements initiated
Accountability and management systems	■ S&OP
	■ Weekly performance review
	■ Project review process
	■ Clear-to-build
	■ Daily walk-around
Documentation	■ All processes linked to transactions and planning documented in standard operating procedure format
Training	■ Utilize documents for training
	■ Audit the processes post-training
Ratchet goals	■ Continue to add and ratchet measurements for continuous process improvement

INDEX